U.S. Marine Corps Summer Survival Course, Training and Skills

Department of Defense

Pentagon
Publishing

UNITED STATES MARINE CORPS

Mountain Warfare Training Center
Bridgeport, CA. 93517-5001

SUMMER SURVIVAL COURSE HANDBOOK

TABLE OF CONTENTS

UNITED STATES MARINE CORPS
Mountain Warfare Training Center
Bridgeport, California 93517-5001

MSVX.02.01
02/06/05

STUDENT HANDOUT

REQUIREMENTS FOR SURVIVAL

TERMINAL LEARNING OBJECTIVE In a survival situation, and given a survival kit, apply the requirements for survival, in accordance with the references. **(MSVX.02.01)**

ENABLING LEARNING OBJECTIVES

(1) Without the aid of references and given the acronym "SURVIVAL", describe in writing the acronym "SURVIVAL", in accordance with the references. **(MSVX.02.01a)**

(2) Without the aid of references, list in writing the survival stressors, in accordance with the references. **(MSVX.02.01b)**

(3) Without the aid of references, list in writing the priorities of work in a survival situation, in accordance with the references. **(MSVX.02.01c)**

(4) Without the aid of references, conduct the priorities of work in a survival situation, in accordance with the references. **(MSVX.02.01d)**

OUTLINE

1. REQUIREMENTS FOR SURVIVAL

a. This positive mental "mind-set" is important in many ways. We usually call it the "will to survive" although you might call it "attitude" as well. This basically means that, if you do not have the right attitude, you may not survive.

b. A guideline that can assist you is the acronym " SURVIVAL". **(MSVX.02.01a)**

(1) Size up.

(a) Size up the situation.

1. Conceal yourself from the enemy.

2. Maintain your wits and use your senses to determine what is happening in your immediate area of influence before making a survival plan.

(b) Size up your surroundings.

1. Determine the rhythm or pattern of the area.

2. Note animal and bird noises and their movement.

3. Note enemy traffic and civilian movement.

(c) Size up your physical condition.

1. Check your wounds and give yourself first aid.

2. Take care to prevent further bodily harm.

3. Evaluate your condition and the condition of your unit prior to developing a plan.

(d) Size up your equipment.

1. Consider how available equipment may affect survival senses; tailor accordingly.

(2) **U**ndue haste makes waste.

(a) Plan your moves so that you can move out quickly without endangering yourself if the enemy is near.

(3) **R**emember where you are.

(a) If you have a map, spot your location and relate it to the surrounding terrain.

(b) Pay close attention to where you are and where you are going. **Constantly orient yourself**.

(c) Try to determine, at a minimum, how your location relates to the following:

1. The location of enemy units and controlled areas.

2. The location of friendly units and controlled areas.

3. The location of local water sources.

4. Areas that will provide good cover and concealment.

(4) __V__anquish fear and panic.

(a) The feeling of fear and panic will be present. The survivor must control these feelings.

(5) __I__mprovise and Improve.

(a) Use tools designed for one purpose for other applications.

(b) Use objects around you for different needs. (i.e. use a rock for a hammer)

(6) __V__alue living.

(a) Place a high value on living.

(b) Refuse to give into the problem and obstacles that face you.

(c) Draw strength from individuals that rise to the occasion.

(7) __A__ct like the natives.

(a) Observe the people in the area to determine their daily eating, sleeping, and drinking routines.

(b) Observe animal life in the area to help you find sources of food and water.

NOTES: Remember that animal reactions can reveal your presence to the enemy. Animals cannot serve as an absolute guide to what you can eat and drink.

(8) __L__ive by your wits, **but for now**, learn basic skills.

(a) Practice basic survival skills during all training programs and exercises.

2. __**STRESS**__. Stress has many positive benefits. Stress provides us with challenges: it gives us chances to learn about our values and strengths. Too much stress leads to distress. While many of these signs may not be self-identified, it remains critical that all survivors remain attentive to each other's signs of distress. Listed are a few common signs of distress found when faced with too much stress:

a. Difficulty in making decisions. (Do not confuse this sign for a symptom of hypothermia).

b. Angry outbursts.
c. Forgetfulness.

d. Low energy level.

e. Constant worrying.

f. Propensity for mistakes.

g. Thoughts about death or suicide.

h. Trouble getting along with others.

i. Withdrawing from others.

j. Hiding from responsibilities.

k. Carelessness.

3. **SURVIVAL STRESSORS. (MSVX.02.01b).** Any event can lead to stress. Often, stressful events occur simultaneously. These events are not stress, but they produce it and are called "stressors". In response to a stressor, the body prepares to either "fight or flight". Stressors add up. Anticipating stressors and developing strategies to cope with them are the two ingredients in the effective management of stress. It is essential that the survivor be aware of the types of stressors they will encounter.

   a. Injury, Illness, or Death. Injury, illness, and death are real possibilities a survivor may face. Perhaps nothing is more stressful than being alone in an unfamiliar environment where you could die from hostile action, an accident, or from eating something lethal.

   b. Uncertainty and Lack of Control. Some people have trouble operating in settings where everything is not clear-cut. This uncertainty and lack of control also add to the stress of being ill, injured or killed.

   c. Environment. A survivor will have to contend with the stressors of weather, terrain and the types of creatures inhabiting an area. Environmental and climactic changes, coupled with insects and animals, are just a few of the challenges awaiting the Marine working to survive.

   d. Hunger and Thirst. Without food and water a person will weaken and eventually die. Getting and preserving food and water take on increasing importance as the length of time in a survival situation increases. With the likelihood of diarrhea, replenishing electrolytes becomes critical. For a Marine used to having his provisions issued, foraging can be a significant source of stress.

e. Fatigue. It is essential that survivors employ all available means to preserve mental and physical strength. While food, water and other energy builders may be in short supply, maximizing sleep to avoid deprivation is a very controllable factor. Further, sleep deprivation directly correlates with increased fear.

f. Isolation. Being in contact with others provides a greater sense of security and a feeling someone is available to help if problems occur.

4. **NATURAL REACTIONS**. Man has been able to survive many changes in his environment throughout the centuries. His ability to adapt physically and mentally to a changing world keeps him alive. The average person will have some psychological reactions in a survival situation. These are some of the major internal reactions you might experience within a survival situation:

   a. Fear. Fear is our emotional response to dangerous situations that we believe have the potential to cause death, injury or illness. Fear can have a positive effect if it forces us to be cautious in situations where recklessness could result in injury.

   b. Anxiety. Anxiety is an uneasy, apprehensive feeling we get when faced with dangerous situations. A survivor reduces his anxiety by performing those tasks that will ensure his coming through the ordeal alive.

   c. Anger and Frustration. Frustration arises when a person is continually thwarted in his attempts to reach a goal. One result of frustration is anger. Getting lost, damaging or forgetting equipment, weather, inhospitable terrain, enemy patrols and physical limitations are just a few sources of frustration and anger. Frustration and anger encourage impulsive reactions, irrational behavior, poorly thought-out decisions, and in some instances, an "I quit" attitude.

   d. Depression. Depression is closely linked with frustration and anger when faced with the privations of survival. A destructive cycle between anger and frustration continues until the person becomes worn down-physically, emotionally and mentally. At this point, he starts to give up, and his focus shifts from "What can I do" to "There is nothing I can do."

   e. Loneliness and Boredom. Man is a social animal and enjoys the company of others. Loneliness and boredom can be another source of depression. Marines must find ways to keep their minds productively occupied.

   f. Guilt. The circumstances leading to your survival situation are sometimes dramatic and tragic. It may be the result of an accident or military action where there was a loss of life. Perhaps you were the only, or one of a few, survivors. While naturally relieved to be alive, you simultaneously may be mourning the deaths of others who were less fortunate. Do not let feelings of guilt prevent you from living.

5. **PRIORITIES OF WORK IN A SURVIVAL SITUATION. (MSVX.02.02c).** Each survival situation will have unique aspects that alter the order in which tasks need to be accomplished. A general guideline is to think in blocks of time.

   a. First 24 hours. The first 24 hours are critical in a survival situation. You must make an initial estimate of the situation. Enemy, weather, terrain, time of day and available resources will determine which tasks need to be accomplished first. They should be the following:

      (1) Shelter.

      (2) Fire.

      (3) Water.

      (4) Signaling.

   b. Second 24 hours. After the first 24 hours have passed, you will now know if you can survive. This time period needs to be spent on expanding your knowledge of the area. By completing the following tasks, you will be able to gain valuable knowledge.

      (1) Tools and weapons. By traveling a short distance from your shelter to locate the necessary resources, you will notice edible food sources and game trails.

      (2) Traps and snares. Moving further away from your shelter to employ traps and snares, you will be able to locate your shelter area from various vantage points. This will enable you to identify likely avenues of approach into your shelter area.

      (3) Pathguards. Knowing the likely avenues of approaches, you can effectively place noise and casualty producing pathguards to ensure the security of your shelter area.

   c. Remainder of your survival situation. This time is spent on continuously improving your survival situation until your rescue.

6. **GROUP SURVIVAL.** Group survival depends largely on the ability to organize activity. An emergency situation does not bring people together for a common goal initially; rather, the more difficult and confusing the situation, the greater are the group's problems.

   a. Groups Morale. High morale must come from internal cohesiveness and not merely through external pressure. The moods and attitudes can become wildly contagious.

Conscious, well-planned organization and leadership on the basis of delegated or shared responsibility often can prevent panic. High group morale has many advantages.

(1) An individual feels strengthened and protected since he realizes that his survival may depend on others whom he trusts.

(2) The group can meet failure with greater persistency.

(3) The group can formulate goals to help each other face the future.

b. Factors that Influence Group Survival. There are numerous factors that will influence whether a group can successfully survive.

(1) Organization of Manpower - Organized action is important to keep all members of the group informed; this way the members of the group will know what to do and when to do it, both under ordinary circumstances and in emergencies.

(2) Selective Use of Personnel - In well-organized groups, the person often does the job that most closely fits his personal qualifications.

(3) Acceptance of Suggestion and Criticisms - The senior man must accept responsibility for the final decision, but must be able to take suggestion and criticisms from others.

(4) Consideration of Time - On-the-spot decisions that must be acted upon immediately usually determine survival success.

(5) Check Equipment - Failure to check equipment can result in failure to survive.

(6) Survival Knowledge and Skills - Confidence in one's ability is increased by acquiring survival knowledge and skills.

REFERENCE:

1. FM 21-76, Survival, 1992.

2. MCRP 3-02h, Survival, Escape, and Evasion, 1999.

3. B-GA-217-001/PT-001, Down but not out, Canadian Survival Guide.

4. AFM 64-5, Search and Rescue Survival, 1969.

UNITED STATES MARINE CORPS
Mountain Warfare Training Center
Bridgeport, California 93517-5001

MSVX.02.02
02/06/05

STUDENT HANDOUT

SURVIVAL KIT

TERMINAL LEARNING OBJECTIVE Without the aid of references, and given an area of operations, construct a personal survival kit, in accordance with the references. **(MSVX.02.02)**

ENABLING LEARNING OBJECTIVES

(1) Without the aid of references, list in writing the components for a survival kit, in accordance with the references. **(MSVX.02.02a)**

(2) Without the aid of references, list in writing one example of each component for a survival kit, in accordance with the references. **(MSVX.02.02b)**

OUTLINE

1. COMPONENTS FOR A SURVIVAL KIT

a. The environment is the key to the types of items you will need in your survival kit. How much equipment you put in your kit depends on how you will carry the kit. A kit on your body will have to be much smaller than one carried in a vehicle.

b. Always layer your survival kit, keeping the most important items on your body.

c. In preparing your survival kit, select items that can be used for more than one purpose.

d. Your survival kit does not need to be elaborate. You only need functional items that will meet your needs and a case to hold them. The case might be a first aid case, an ammunition pouch, or another suitable case. This case should be-

(1) Water repellent or waterproof.

(2) Easy to carry or attach to your body.

(3) Suitable to accept various sized items.

(4) Durable.

e. When constructing a survival kit, you should have the following components: **(MSVX.02.02a)**

(1) Fire starting items.

(2) Water procurement items.

(3) Food procurement items.

(4) Signaling items.

(5) First aid items.

(6) Shelter items.

2. **<u>ITEMS CONTAINED WITHIN EACH COMPONENT</u> (MSV.02.02b*)***

a. <u>Fire Starting Items</u>.

(1) Matches.

(2) Magnifying glass.

(3) Flint and Steel.

(4) Lighter.

(5) Potassium Permanganate, with a container of sugar or anti-freeze.

(6) Prepackaged Tinder.
-Commercially Manufactured
-Cotton Balls and Petroleum Jelly

<u>b. Water Procurement Items</u>

(1) Water Disinfecting Chemicals.
-Iodine Tablets
-Betadine Solution
-Iodine Solution

(2) Metal Container. (Serves for boiling water)
-Canteen Cup
-Survival Kit Container
-Any Suitable can that contained no petroleum products.

(3) Water Carrying Items.
-Canteen
-Plastic Bag
-Plastic/Metal/Glass Container which contained no petroleum products.

c. Food Procurement Items

(1) Fish.
-Various sized hooks
-Various sized sinkers/weights
-Metal leaders and swivels
-Small weighted jigs
-Fishing line
-Think about the size of fish for that environment when selecting weights and sizes.

(2) Game.
-Snares
*Commercially Manufactured
*Aircraft Cable
*Tie Wire
-Bait
*MRE Cheese Spread or Peanut Butter Package
-550 Cord for Gill Net and Trap Construction
-Engineer/Marking Tape
-Sling shot rubber and pouch

d. Signaling Items

(1) Day.
-Mirror
-Whistle
-Pyrotechnics (Smoke, Pen Flares)
-Air Panels

(2) Night.
-Pyrotechnics (Pen Flares, Star Clusters)
-Lights (Flashlight, Strobe, Chemlight)
-Whistle

e. Shelter Items

(1) Cordage.
-550 Cord.
-Wire.
-Communication wire
-Tie wire

(2) Finger Saw.

(3) Sewing Kit with Needles for construction/repair of clothing.

(4) Tentage.
-poncho
-tarp
-space blanket
-plastic trash bags

f. First Aid Items

(1) Band-Aids.
-Steristrips
-Adhesive Tape
-Non-stick pads, 4x4's, Gauze, Battle Dressings
-Muslin Bandage

(2) Ointments.
-Burn
-Anti-septic

(3) Miscellaneous.
-Salt
-Sugar
-Eye Wash
-Alcohol prep pads
-Suture Kit
-Scalpel
-Vile of Yarrow

g. Miscellaneous items.

(1) Fingernail clippers.

(2) Compass.

(3) Notebook with pen or pencil.

(4) Wood eye screws and nails.

(5) Surgical tubing.

Note: It is assumed that the Marine is always carrying a high quality fixed bladed knife, a multi-tool knife, and a sharpening stone.

REFERENCE:

1. FM 21-76, Survival, 1992.

2. Barry Davies BME, SAS Escape Evasion and Survival Manual, 1996.

3. John Wiesman, SAS Survival Guide, 1986.

UNITED STATES MARINE CORPS
Mountain Warfare Training Center
Bridgeport, California 93517-5001

MSVX.02.03
2/6/05

STUDENT HANDOUT

WATER PROCUREMENT

TERMINAL LEARNING OBJECTIVE In a survival situation, and given a survival kit, and water procurement materials, obtain potable water, in accordance with the references. **(MSVX.02.03)**

ENABLING LEARNING OBJECTIVES

(1) Without the aid of references, list in writing the types of incidental water, in accordance with the references. **(MSVX.02.03a)**

(2) Without the aid of references, list in writing the hazardous fluids to avoid substituting for potable water, in accordance with the references. **(MSVX.02.03b)**

(3) Without the aid of references, list in writing the methods for disinfecting water, in accordance with the references. **(MSVX.02.03c)**

(4) Without the aid of references and given a military bottle of water purification tablets, state in writing its self-life, in accordance with the references. **(MSVX.02.03d)**

(5) Without the aid of references, and given the water temperature and chemical concentration, state in writing the contact time, in accordance with the references. **(MSVX.02.03e)**

(6) Without the aid of references, construct a solar still , in accordance with the references. **(MSVX.02.03f)**

OUTLINE

1. WATER INTAKE

a. Thirst is not a strong enough sensation to determine how much water you need.

b. The best plan is to drink, utilizing the OVER DRINK method. Drink plenty of water anytime it is available and particularly when eating.

c. Dehydration is a major threat. A loss of only 5 % of your body fluids causes thirst, irritability, nausea, and weakness; a 10% loss causes dizziness, headache, inability to walk, and a tingling sensation in limbs; a 15% loss causes dim vision, painful urination, swollen tongue, deafness, and a feeling of numbness in the skin; also a loss of more than 15% body fluids could result in death.

d. Your water requirements will be increased if:

(1) You have a fever.

(2) You are experiencing fear or anxiety.

(3) You evaporate more body fluid than necessary. (i.e., not using the proper shelter to your advantage)

(4) You have improper clothing.

(5) You ration water.

(6) You overwork.

2. **INCIDENTAL WATER. (MSVX.02.03a)**

a. During movement, you may have to ration water until you reach a reliable water source. Incidental water may sometimes provide opportunities to acquire water. Although not a reliable or replenished source, it may serve to stretch your water supply or keep you going in an emergency. The following are sources for incidental water:

(1) Dew. In areas with moderate to heavy dew, dew can be collected by tying rags or tuffs of fine grass around your ankles. While walking through dewy grass before sunrise, the rags or grass will saturate and can be rung out into a container. The rags or grass can be replaced and the process is repeated.

(2) Rainfall. Rainwater collected directly in clean container or in plants that contain no harmful toxins is generally safe to drink without disinfecting. The survivor should always be prepared to collect rainfall at a moments notice. An inverted poncho works well to collect rainfall.

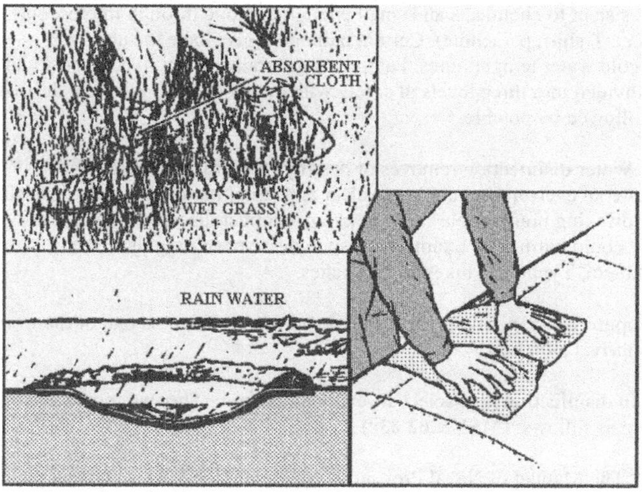

3. **HAZARDOUS FLUIDS** **(MSVX.02.03b*)***

   a. Survivors have occasionally attempted to augment their water supply with other fluids, such as alcoholic beverages, urine, blood, or seawater. While it is true that each of these fluids has a high water content, the impurities they contain may require the body to expend more fluid to purify them. Some hazardous fluids are:

      (1) Sea water. Sea water in more than minimal quantities is actually toxic. The concentration of sodium and magnesium salts is so high that fluid must be drawn from the body to eliminate the salts and eventually the kidneys cease to function.

      (2) Alcohol. Alcohol dehydrates the body and clouds judgment. Super-cooled liquid, if ingested, can cause immediate frostbite of the throat, and potential death.

      (3) Blood. Blood, besides being salty, is a food. Drinking it will require the body to expend additional fluid to digest it.

      (4) Urine. Drinking urine is not only foolish, but also dangerous. Urine is nothing more than the body's waste. Drinking it only places this waste back into the body, which requires more fluid to process it again.

4. **WATER QUALITY**. Water contains minerals, toxins, and pathogens. Some of these, consumed in large enough quantities may be harmful to human health. Pathogens are our primary concern. Pathogens are divided into Virus, Cysts, Bacteria, and Parasites. Certain

pathogens are more resistant to chemicals and small enough to move through microscopic holes in equipment (i.e., T-shirt, parachute). Certain pathogens also have the ability to survive in extremely cold water temperatures. Pathogens generally do not live in snow and ice. Water quality is divided into three levels of safety with disinfection as the most desired level, then purified, followed by potable.

a. Disinfection. Water disinfection removes or destroys harmful microorganisms. Giardia cysts are an ever-present danger in clear appearing mountain water throughout the world. By drinking non-potable water you may contract diseases or swallow organisms that could harm you. Examples of such diseases or organisms are: Dysentery, Cholera, Typhoid, Flukes, and Leeches.

b. Remember, impure water, no matter how overpowering the thirst, is one of the worst hazards in a survival situation.

c. The first step in disinfecting is to select a treatment method. The two methods we will discuss are as follows: **(MSVX.02.03c)**

(1) Heat. The Manual of Naval Preventive Medicine (P-5010) states that you must bring the water to a rolling boil before it is considered safe for human consumption. This is the most preferred method.

(a) Bringing water to the boiling point will kill 99.9% of all Giardia cysts. The Giardia cyst dies at 60°C and Cryptosporidium dies at 65C. Water will boil at 14,000' at 86°C and at 10,000' at 90C. With this in mind you should note that altitude does not make a difference unless you are extremely high.

(2) Chemicals. There are numerous types of chemicals that can disinfect water. Below are a few of the most common. In a survival situation, you will use whatever you have available.

(a) Iodine Tablets.

(b) Chlorine Bleach.

(c) Iodine Solution.

(d) Betadine Solution.

(e) Military water purification tablets. **(MSVX.02.03d)** These tablets are standard issue for all DOD agencies. These tablets have a shelf-life of four years from the date of manufacture, unless opened. Once the seal is broken, they have a shelf-life of one year, not to exceed the initial expiration date of four years.

49703
Month / Year / Batch Number

(3) Water Disinfection Techniques and Halogen Doses.

Iodination techniques Added to 1 liter or quart of water	Amount for 4 ppm	Amount for 8 ppm
Iodine tablets Tetraglcine hydroperiodide EDWGT Potable Aqua Globaline	½ tablet	1 tablet
2% iodine solution (tincture)	0.2 ml 5 gtts	0.4 ml 10 gtts
10% povidone-iodine solution*	0.35 ml 8 gtts	0.70 ml 16 gtts
Chlorination techniques	Amount for 5 ppm	Amount for 10 ppm
Household bleach 5% Sodium hypochlorie	0.1 Ml 2 gtts	0.2 ml 4 gtts
AquaClear Sodium dichloroisocyanurate		1 tablet
AquaCure, AquaPure, Chlor-floc Chlorine plus flocculating agent		8 ppm 1 tablet

*Providone-iodine solutions release free iodine in levels adequate for disinfection, but scant data is available.

Measure with dropper (1 drop=0.05 ml) or tuberculin syringe

Ppm-part per million gtts-drops ml-milliliter

Concentration of Contact time in minutes at various water temperatures (WSVX.02.03e) halogen

	5 C / 40 F	15 C / 60 F	30 C / 85 F
2 ppm	240	180	60
4 ppm	180	50	45
8 ppm	60	30	15

Note: These contact times have been extended from the usual recommendations to account for recent data that prolonged contact time is needed in very cold water to kill *Giardia* cysts.

Note: chemicals may not destroy Cryptosporidium.

d. Purification. Water purification is the removal of organic and inorganic chemicals and particulate matter, including radioactive particles. While purification can eliminate offensive color, taste, and odor, it may not remove or kill microorganisms.

(1) Filtration. Filtration purifying is a process by which commercial manufacturers build water filters. The water filter is a three tier system. The first layer, or grass layer, removes the larger impurities. The second layer, or sand layer, removes the smaller impurities. The final layer, or charcoal layer (**not the ash but charcoal from a fire**), bonds and holds the toxins. All layers are placed on some type of straining device and the charcoal layer should be at least 5-6 inches thick. Layers should be changed frequently and straining material should be boiled. Remember, this is not a disinfecting method, cysts can possibly move through this system.

WATER FILTER

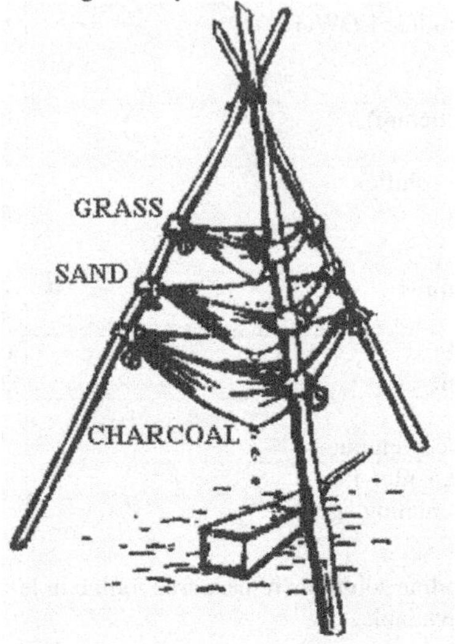

(2) Commercial Water Filters. Commercial water filters are generally available in most retail stores and may be with you. Understanding what the filter can do is the first step in safeguarding against future illnesses.

(a) A filter that has a .3 micron opening or larger will not stop Cryptosporidium.

(b) A filter system that does not release a chemical (i.e., iodine) may not kill all pathogens.

(c) A filter that has been overused may be clogged. Usage may result in excessive pumping pressure that can move harmful pathogens through the opening.

e. Potable. Potable indicates only that a water source, on average over a period of time, contains a "minimal microbial hazard," so the statistical likelihood of illness is acceptable.

(1) Sedimentation. Sedimentation is the separation of suspended particles large enough to settle rapidly by gravity. The time required depends on the size of the particle. Generally, 1 hour is adequate if the water is allowed to sit without agitation. After sediment has formed on the bottom of the container, the clear water is decanted or filtered from the top. Microorganisms, especially cysts, eventually settle, but this takes longer and the organisms are easily disturbed during pouring or filtering. Sedimentation should not be considered a means of disinfection and should be used only as a last resort or in an extreme tactical situation.

5. **SOLAR STILLS. (MSVX.2.3f)**

a. Solar stills are designed to supplement water reserves. Contrary to belief, they will not provide enough water to meet the daily requirement for water.

b. Above-Ground Solar Still. This device allows the survivor to make water from vegetation. To make the aboveground solar still, locate a sunny slope on which to place the still, a clear plastic bag, green leafy vegetation, and a small rock.

(1) Construction.

(a) Fill the bag with air by turning the opening into the breeze or by "scooping" air into the bag.

(b) Fill the bag half to three-quarters full of green leafy vegetation. Be sure to remove all hard sticks or sharp spines that might puncture the bag.

CAUTION
Do not use poisonous vegetation. It will provide poisonous liquid.

(c) Place a small rock or similar item in the bag.

(d) Close the bag and tie the mouth securely as close to the end of the bag as possible to keep the maximum amount of air space. If you have a small piece of tubing, small straw, or hollow reed, insert one end in the mouth of the bag before tying it securely. Tie off or plug the tubing so that air will not escape. This tubing will allow you to drain out condensed water without untying the bag.

(e) Place the bag, mouth downhill, on a slope in full sunlight. Position the mouth of the bag slightly higher than the low point in the bag.

(f) Settle the bag in place so that the rock works itself into the low point in the bag.

(g) To get the condensed water from the still, loosen the tie and tip the bag so that the collected water will drain out. Retie the mouth and reposition the still to allow further condensation.

(h) Change vegetation in the bag after extracting most of the water from it.

(i) Using 1 gallon zip-loc bag instead of trash bags is a more efficient means of construction.

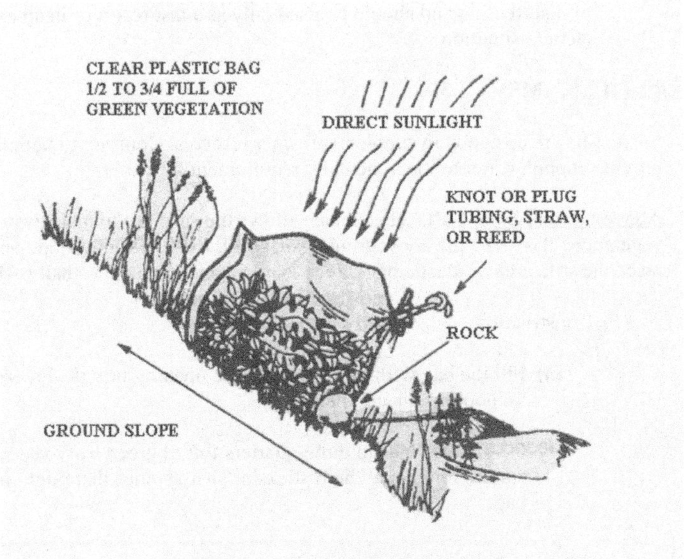

ABOVE GROUND SOLAR STILL

c. Below-Ground Solar Still. Materials consist of a digging stick, clear plastic sheet, container, rock, and a drinking tube.

(1) Construction.

(a) Select a site where you believe the soil will contain moisture (such as a dry streambed or a low spot where rainwater has collected). The soil should be easy to dig, and will be exposed to sunlight.

(b) Dig a bowl-shaped hole about 1 meter across and 24 inches deep.

(c) Dig a sump in the center of the hole. The sump depth and perimeter will depend on the size of the container you have to place in it. The bottom of the sump should allow the container to stand upright.

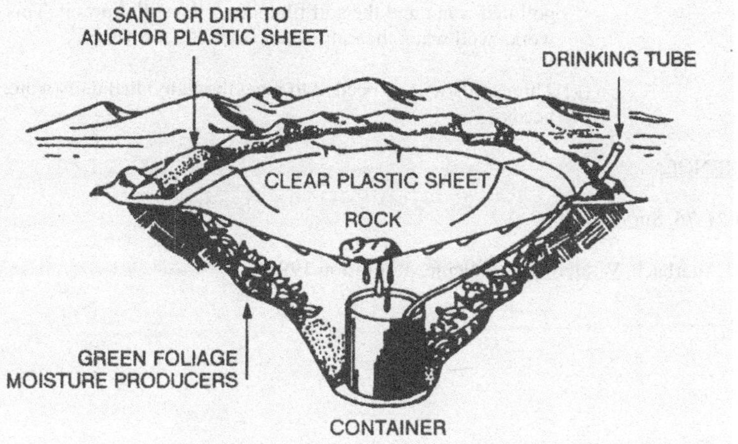

<u>BELOW GROUND SOLAR STILL</u>

(d) Anchor the tubing to the container's bottom by forming a loose overhand knot in the tubing. Extend the unanchored end of the tubing up, over, and beyond the lip of the hole.

(e) Place the plastic sheet over the hole, covering its edges with soil to hold in place. Place a rock in the center of the plastic sheet.

(f) Lower the plastic sheet into the hole until it is about 18 inches below ground level. Make sure the cone's apex is directly over the container. Ensure the plastic does not touch the sides of the hole because the earth will absorb the moisture.

(g) Put more soil on the edges of the plastic to hold it securely and prevent the loss of moisture.

(h) Plug the tube when not in use so that moisture will not evaporate.

(i) Plants can be placed in the hole as a moisture source. If so, dig out additional soil from the sides.

(j) If polluted water is the only moisture source, dig a small trough outside the hole about 10 inches from the still's lip. Dig the trough about 10 inches deep and 3 inches wide. Pour the polluted water in the trough. Ensure you do not spill any polluted water around the rim of the hole where the plastic touches the soil. The trough holds the polluted water and the soil filters it as the still draws it. This process works well when the only water source is salt water.

(k) Three stills will be needed to meet the individual daily water intake needs.

<u>REFERENCE</u>:

1. FM 21-76, <u>Survival</u>, 1992.

2. Paul Auerbach, <u>Wilderness Medicine</u>, 3rd Edition 1995.

UNITED STATES MARINE CORPS
Mountain Warfare Training Center
Bridgeport, California 93517-5001

MSVX.02.04
2/6/05

STUDENT HANDOUT

EXPEDIENT SHELTERS AND FIRES

TERMINAL LEARNING OBJECTIVE.

(1) In a survival situation, and given a survival kit, construct an expedient survival shelter, in accordance with the references. **(MSVX.02.04)**

(2) In a survival situation, and given a survival kit, construct survival fires utilizing man made and natural materials, in accordance with the references. **(MSVX.02.05)**

ENABLING LEARNING OBJECTIVES

(1) Without the aid of references, list in writing the characteristics of a safe expedient shelter, in accordance with the references. **(MSVX.02.04a)**

(2) Without the aid of references, list in writing the hazards to avoid when using natural shelters, in accordance with the references. **(MSVX.02.04b)**

(3) Without the aid of references, list in writing the types of man-made survival shelters, in accordance with the references. **(MSVX.02.04c)**

(4) Without the aid of references, list in writing the tactical fire lay, in accordance with the references. **(MSVX.02.05a)**

(5) Without the aid of references, list in writing the materials utilized to construct survival fire, in accordance with the references. **(MSVX.02.05b)**

(6) Without the aid of references, start a fire using a primitive method, in accordance with the references. **(MSVX.02.05c)**

(7) Without the aid of references, start a fire using man made materials, in accordance with the references. **(MSVX.02.05d)**

OUTLINE

1. **BASIC CHARACTERISTICS FOR SHELTER**. **(MSVX.02.04a)** Any type of shelter, whether it is a permanent building, tentage, or an expedient shelter must meet six basic criteria to be safe and effective. The characteristics are:

   a. Protection From the Elements. The shelter must provide protection from rain, snow, wind, sun, etc.

   b. Heat Retention. It must have some type of insulation to retain heat; thus preventing the waste of fuel.

   c. Ventilation. Ventilation must be constructed, especially if burning fuel for heat. This prevents the accumulation of carbon monoxide. Ventilation is also needed for carbon dioxide given off when breathing.

   d. Drying Facility. A drying facility must be constructed to dry wet clothes.

   e. Free from Natural Hazards. Shelters should not be built in areas of avalanche hazards, under rock fall or "standing dead" trees have the potential to fall on your shelter.

   f. Stable. Shelters must be constructed to withstand the pressures exerted by severe weather.

2. **NATURAL SHELTERS**. Natural shelters are usually the preferred types because they take less time and materials construct. The following may be made into natural shelters with some modification. **(MSVX.02.04b)**

   a. Caves or Rock Overhangs. Can be modified by laying walls of rocks, logs or branches across the open sides.

   b. Hollow Logs. Can be cleaned or dug out, then enhanced with ponchos, tarps or parachutes hung across the openings.

   c. Hazards of Natural Shelters.

      (1) Animals. Natural shelters may already be inhabited (i.e. bears, coyotes, lions, rats, snakes, etc.). Other concerns from animals may be disease from scat or decaying carcasses.

      (2) Lack of Ventilation. Natural shelters may not have adequate ventilation. Fires may be built inside for heating or cooking but may be uncomfortable or even dangerous because of the smoke build up.

(3) Gas Pockets. Many caves in a mountainous region may have natural gas pockets in them.

(4) Instability. Natural shelters may appear stable, but in reality may be a trap waiting to collapse.

3. **MAN-MADE SHELTERS**. **(MSVX.02.04c)** Many configurations of man-made shelters may be used. Over-looked man-made structures found in urban or rural environments may also provide shelter (i.e. houses, sheds, or barns). Limited by imagination and materials available, the following man-made shelters can be used in any situation.

a. Poncho Shelter

b. Sapling Shelter.

c. Lean-To.

d. Double Lean-To.

e. A-frame Shelter.

f. Fallen Tree Bivouac.

4. **CONSTRUCTION OF MAN-MADE SHELTERS**. To maximize the shelter's effectiveness, Marines should take into consideration the following prior to construction.

a. Considerations.

(1) Group size.

(2) Low silhouette and reduced living area dimensions for improved heat conservation.

(3) Avoid exposed hill tops, valley floors, moist ground, and avalanche paths.

(4) Create a thermal shelter by applying snow, if available, to roof and sides of shelter.

(5) Location of site to fire wood, water, and signaling, if necessary.

(6) How much time and effort needed to build the shelter.

(7) Can the shelter adequately protect you from the elements (sun , wind, rain, and snow). Plan on worst case scenario.

(8) Are the tools available to build it. If not, can you make improvised tools?
(9) Type and amount of materials available to build it.

(10) When in a tactical environment, you must consider the following:

(a) Provide concealment from enemy observation.

(b) Maintain camouflaged escape routes.

(c) Use the acronym BLISS as a guide.

B – Blend in with the surroundings.
L – Low silhouette.
I – Irregular shape.
S – Small.
S – Secluded located.

b. <u>Poncho Shelter</u>. This is one of the easiest shelters to construct. Materials needed for construction are cord and any water-repellent material (i.e. poncho, parachute, tarp). It should be one of the first types of shelter considered if planning a short stay in any one place.

(1) Find the center of the water-repellent material by folding it in half along its long axis.

(2) Suspend the center points of the two ends using cordage.

(3) Stake the four corners down, with sticks or rocks.

c. <u>Sapling Shelter</u>. This type of shelter is constructed in an area where an abundance of saplings are growing. It is an excellent evasion shelter.

(1) Find or clear an area so that you have two parallel rows of saplings at least 4' long and approximately 1 ½' to 2' apart.

(2) Bend the saplings together and tie them to form several hoops which will form the framework of the shelter.

(3) Cover the hoop with a water-repellent covering.

(4) The shelter then may be insulated with leaves, brush, snow, or boughs.

(5) Close one end permanently. Hang material over the other end to form a door.

d. Lean-To. A lean-to is built in heavily forested areas. It requires a limited amount of cordage to construct. The lean-to is an effective shelter but does not offer a great degree of protection from the elements.

(1) Select a site with two trees (4-12" in diameter), spaced far enough apart that a man can lay down between them. Two sturdy poles can be substituted by inserting them into the ground the proper distance apart.

(2) Cut a pole to support the roof. It should be at least 3-4" in diameter and long enough to extend 4-6" past both trees. Tie the pole horizontally between the two trees, approximately 1 meter off the deck.

(3) Cut several long poles to be used as stringers. They are placed along the horizontal support bar approximately every 1 ½' and laid on the ground. All stringers may be tied to or laid on the horizontal support bar. A short wall or rocks or logs may be constructed on the ground to lift the stringers off the ground, creating additional height and living room dimensions.

(4) Cut several saplings and weave them horizontally between the stringers. Cover the roof with water-repellent and insulating material.

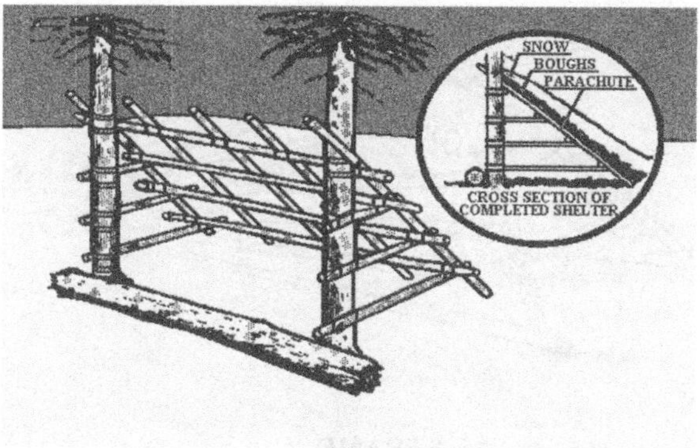

LEAN-TO

e. Double Lean-To. The double lean-to shelter is constructed for 2-5 individuals. It is constructed by making two lean-to's and placing them together.

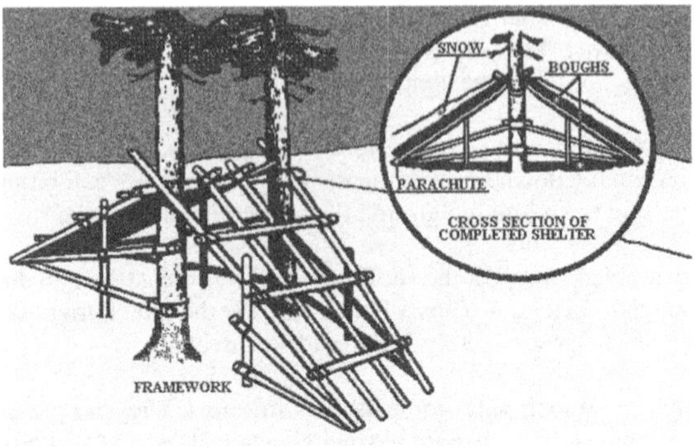

DOUBLE LEAN-TO

f. A-Frame Shelter. An A-Frame shelter is constructed for 1-3 individuals. After the frame work is constructed, bough/tentage is interwoven onto the frame and snow, if available, is packed onto the outside for insulation.

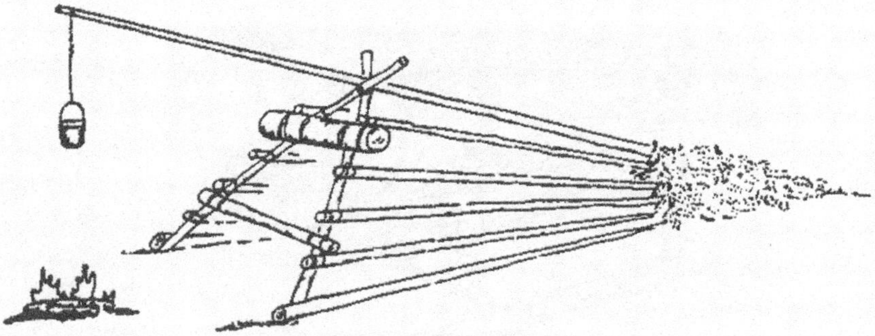

A-FRAME

g. Fallen Tree Bivouac. The fallen tree bivouac is an excellent shelter because most of the work has already been done.

(1) Ensure the tree is stable prior to constructing.

(2) Branches on the underside are cut away making a hollow underneath.

(3) Place additional insulating material to the top and sides of the tree.

(4) A small fire is built outside of the shelter.

FALLEN TREE BIVOUAC

5. **REFLECTOR WALLS.** Heating a shelter requires a slow fire that produces lots of steady heat over a long period of time. A reflector wall should be constructed for all open ended shelters. A reflector wall is constructed with a flat rock or a stack of green logs propped behind the fire. A surprising amount of heat will bounce back from the fire into the shelter.

6. **FIRES**. Fires fall into two main categories: those built for cooking and those built for warmth and signaling. The basic steps are the same for both: preparing the fire lay, gathering fuel, building the fire, and properly extinguishing the fire.

   a. Preparing the fire lay. There are two types of fire lays: fire pit and Dakota hole. Fire pits are probably the most common.

      (1) Create a windbreak to confine the heat and prevent the wind from scattering sparks. Place rocks or logs used in constructing the fire lay parallel to the wind. The prevailing downwind end should be narrower to create a chimney effect.

      (2) Avoid using wet rocks. Heat acting on the dampness in sandstone, shale, and stones from streams may cause them to explode.

      (3) Dakota Hole. **(MSVX.02.05a)** The Dakota Hole is a tactical fire lay. Although no fire is 100% tactical, this fire lay will accomplish certain things:

         (a) Reduces the signature of the fire by placing it below ground.

         (b) Provides more of a concentrated heat source to boil and cook, thus preserving fuel and lessening the amount of burning time.

         (c) By creating a large air draft, the fire will burn with less smoke than the fire pit.

(d) It is easier to light in high winds.

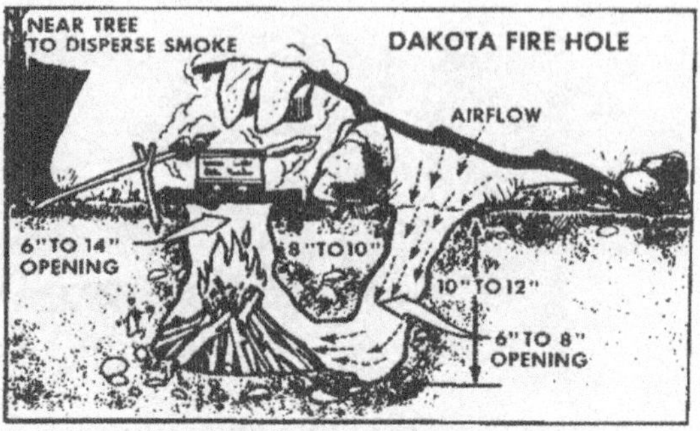

DAKATA HOLE

b. Gather Fuel. Many Marines take shortcuts when gathering firewood. Taking a few extra minutes can mean the difference between ease and frustration when building a fire. **(MSVX.02.05b)**

(1) Tinder. Tinder is the initial fuel. It should be fine and dry. Gather a double handful of tinder for the fire to be built and an extra double handful to be stored in a dry place for the following morning. Dew can moisten tinder enough to make lighting the fire difficult. Some examples are:

(a) Shredded cedar/juniper bark, pine needles.

(b) Dry grass.

(c) Slivers shaved from a dry stick.

(d) Hornet's nest.

(e) Natural fibers from equipment supplemented with pine pitch (i.e., cotton battle dressing).

(f) Cotton balls and petroleum jelly or Char-cloth.

Note: Sticks used for tinder should be dry and not larger than the diameter of a toothpick.

(2) Kindling. This is the material that is ignited by the tinder that will burn long enough to ignite the fuel.

(a) Small sticks/twigs pencil-thick up to the thickness of the thumb. Ensure that they are dry.

(b) Due to a typically large resin content, evergreen limbs often make the best kindling. They burn hot and fast, but typically do not last long.

(3) Fuel Wood. Fuel Wood is used to keep the blaze going long enough to fulfill its purpose. Ideally, it should burn slow enough to conserve the wood pile, make plenty of heat, and leave an ample supply of long-lasting coals.

(a) Firewood broken from the dead limbs of standing trees or windfalls held off the ground will have absorbed less moisture and therefore should burn easily.

(b) Refrain from cutting down live, green trees.

(c) Softwoods (evergreens and conifers) will burn hot and fast with lots of smoke and spark, leaving little in the way of coals. Hardwoods (broad leaf trees) will burn slower with less smoke and leave a good bed of coals.

(d) Learn the woods indigenous to the area. Birch, dogwood, and maple are excellent fuels. Osage orange, ironwood, and manzanita, though difficult to break up, make terrific coals. Aspen and cottonwood burn clean but leave little coals.

(e) Stack your wood supply close enough to be handy, but far enough from the flames to be safe. Protect your supply from additional precipitation.

(f) If you happen to go down in an aircraft that has not burned, a mixture of gas and oil may be used. Use caution when igniting this mixture.

c. Building the Fire. The type of fire built will be dependent upon its intended use; either cooking or heating and signaling.

(1) Cooking Fires. The following listed fires are best used for cooking:

(a) Teepee Fire. The teepee fire is used to produce a concentrated heat source, primarily for cooking. Once a good supply of coals can be seen, collapse the teepee and push embers into a compact bed.

TEEPEE FIRE

(2) Heating and Signaling Fires.

(a) Pyramid Fire. Pyramid fires are used to produce large amounts of light and heat. They will dry out wet wood or clothing.

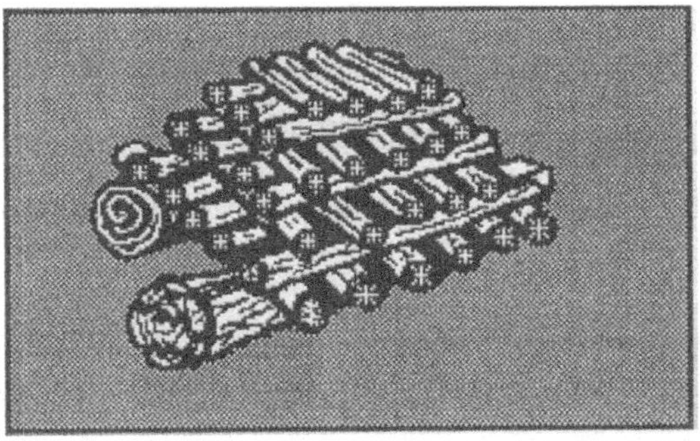

PYRAMID FIRE

d. Starting Fires. Starting a fire is done by a source of ignition and falls into two categories; modern igniters and primitive methods.

(1) Modern Methods. Modern igniters use modern devices we normally think of to start a fire. Reliance upon these methods may result in failure during a survival situation. These items may fail when required to serve their purpose.

(a) Matches and Lighters. Ensure you waterproof these items.

(b) Convex Lens. Binocular, camera, telescopic sights, or magnifying lens are used on bright, sunny days to ignite tinder.

(c) Flint and Steel. Sometimes known as metal matches or "Mag Block". Scrape your knife or carbon steel against the flint to produce a spark onto the tinder. Some types of flint & steel designs will have a block of magnesium attached to the device which can be shaved onto the tinder prior to igniting. Other designs may have magnesium mixed into the flint to produce a higher quality of spark.

(2) Primitive Methods. Primitive fire methods are those developed by early man. There are numerous techniques that fall into this category. The only method that will be taught at MCMWTC is the Bow & Drill.

(3) Bow & Drill. **(MSVX.02.05b)** The technique of starting a fire with a bow & drill is a true field expedient fire starting method which requires a piece of cord and knife from your survival kit to construct. The components of the bow & drill are bow, drill, socket, fire board, ember patch, and birds nest.

(a) Bow. The bow is a resilient, green stick about 3/4 of an inch in diameter and 30-36 inches in length. The bow string can be any type of cord, however, 550 cord works best. Tie the string from one end of the bow to the other, without any slack.

(b) Drill. The drill should be a straight, seasoned hardwood stick about 1/2 to 3/4 of an inch in diameter and 8 to 12 inches in length. The top end is tapered to a blunt point to reduce friction generated in the socket. The bottom end is slightly rounded to fit snugly into the depression on the fire board.

(c) Socket. The socket is an easily grasped stone or piece of hardwood or bone with a slight depression on one side. Use it to hold the drill in place and to apply downward pressure.

(d) Fire board. The fire board is a seasoned softwood board which should ideally be 3/4 of an inch thick, 2-4 inches wide, and 8-10 inches long. Cut a depression 3/4 of an inch from the edge on one side of the fire board. Cut a U-shape notch from the edge of the fire board into the depression. This notch is designed to collect and form an ember which will be used to ignite the tinder.

(e) Ember Patch. The ember patch is made from any type of suitable material (i.e., leather, aluminum foil, bark). It is used to catch and transfer the ember from the fire board to the birds nest. Ideally, it should be 4 inches by 4 inches in size.

(f) Birds Nest. The birds nest is a double handful of tinder which will be made into the shape of a nest. Tinder must be dry and finely shredded material (i.e., outer bark from juniper/cedar/sage brush or inner bark from cottonwood/aspen or dry grass/moss). Lay your tinder out in two equal rows about 4 inches wide and 8-12 inches long. Loosely roll the first row into a ball and knead the tinder to further break down the fibers. Place this ball perpendicular onto the second row of tinder and wrap. Knead the tinder until all fibers of the ball are interwoven. Insert the drill half way into the ball to form a partial cylinder. This is where the ember will be placed.

(4) Producing a fire using the bow & drill.

(a) Place the ember patch under the U-shaped notch.

(b) Assume the kneeling position, with the left foot on the fire board near the depression.

(c) Load the bow with the drill. Ensure the drill is between the wood of the bow and bow string. Place the drill into the depression on the fire board. Place the socket on the tapered end of the drill.

(d) Use the left hand to hold the socket while applying downward pressure.

(e) Use the right hand to grasp the bow. With a smooth sawing motion, move the bow back and forth to twirl the drill.

(f) Once you have established a smooth motion, smoke will appear. Once smoke appears, apply more downward pressure and saw the bow faster.

(g) When a thick layer of smoke has accumulated around the depression, stop all movement. Remove the bow, drill, and socket from the fire board, without moving the fire board. Carefully remove your left foot off the fire board.

(h) Gently tap the fire board to ensure all of the ember has fallen out of the U-shaped notch and is lying on the ember patch. Remove the fire board.

(i) Slowly fan the black powder to solidify it into a glowing ember. Grasping the ember patch, carefully drop the ember into the cylinder of the birds nest.

(j) Grasp the birds nest with the cylinder facing towards you and parallel to the ground. Gently blow air into the cylinder. As smoke from the nest becomes thicker, continue to blow air into the cylinder until fire appears.

(5) Trouble Shooting the Bow & Drill

(a) Drill will not stay in depression- Apply more downward pressure and/or increase width/depth of depression.

(b) Drill will not twirl- Lessen the amount of downward pressure and/or tighten bow string.

(c) Socket smoking- Lessen the amount of downward pressure. Wood too soft when compared to hardness of drill. Add some lubrication: animal fat, oil, or grease.

(d) No smoke- Drill and fire board are the same wood. Wood may not be seasoned. Check drill to ensure that it is straight. Keep left hand locked against left shin while sawing.

(e) Smoke but no ember- U-shaped notch not cut into center of the depression.

(f) Bow string runs up and down drill- Use a locked right arm when sawing. Check drill to ensure that it is straight. Ensure bow string runs over the top of the left boot.

(g) Birds nest will not ignite- Tinder not dry. Nest woven too tight. Tinder not kneaded enough. Blowing too hard (ember will fracture).

e. Extinguishing the Fire. The fire must be properly extinguished. This is accomplished by using the drown, stir, and feel method.

(1) Drown the fire by pouring at water in the fire lay.

(2) Stir the ember bed to ensure that the fire is completely out.

(3) Check the bed of your fire by feeling for any hot spots.

(4) If any hot spots are found, start the process all over again.

REFERENCE:

1. FM 21-76, Survival, 1992.

2. Chris Janowski, A Manual that Could Save your Life, 1996.

3. John Wiesman, SAS Survival Guide, 1993.

4. AFP 36-2246, Aircrew Survival, 1996.

UNITED STATES MARINE CORPS

Mountain Warfare Training Center
Bridgeport, California 93517-5001

MSVX.01.01
2/6/05

STUDENT HANDOUT

CORE VALUES AND MOUNTAIN LEADERSHIP CHALLENGES

LESSON PURPOSE: The purpose of this period of instruction is to emphasize the vital role of leadership in the conduct of successful operations and to promote among leaders at all levels an understanding of the problems common to units operating in a summer mountainous environment. This lesson relates to all of the training that you will receive here at MCMWTC.

OUTLINE

1. **POSITIVE LEADERSHIP AND THE RIGHT ATTITUDE**. Leadership must be by example. At first, harsh and unfamiliar conditions tend to be frightening. Marines will find themselves up against many challenges they have never met before and the environment will constantly remind them that they can become a casualty if they make mistakes. Aggressive leadership, which consistently meets and overcomes the challenges of the environment, is essential to mission accomplishment. There will be two enemies to contend with: the enemy soldier and the environment itself. The first step towards defeating these enemies is getting your Marines in the right mental attitude. The leader must maintain a positive attitude towards the mission, his Marines, and the equipment they have to carry out the job. You can be defeated psychologically, if you are not aware of the symptoms of a poorly motivated unit.

    a. Core Values applied in a survival situation. Core value are very applicable to leadership in a survival situation. In the absence of constant adherence to our core values, subordinate Marines may suffer the consequences of injury and in extreme cases, DEATH.

    b. Honor. Honor is integrity, responsibility, and accountability. Without honor in this type of environment, decisions are made with inaccurate and/or misleading information. Dishonorable actions or intentions ultimately lead to unethical behavior...trust between Marines will erode, dependability will become unimportant, and the resulting consequences to energy conservation will greatly affect a unit's cohesion and ability to accomplish its mission.

    c. Courage. Courage is doing the right thing for the right reason. Courage in this type of environment for small unit leaders to take charge of their unit. The added stress and pressure induced in mountainous terrain makes the tough decisions even tougher to make: **courageous leaders take that extra step.** Trust and confidence in leadership is built upon thorough training to ensure their men are capable of overcoming the fear and uncertainty of enemy and the harsh environment. A few examples of courage are:

        (1) Ensuring Marines have the proper clothing on for the task they are performing.

(2) Ensuring Marines are maintaining security, regardless of weather conditions.

(3) Preventing lethargy and laziness from affecting a marine's decision making skills.

(4) Rising to the challenges of a harsh environment, and beating the cold weather with solid leadership.

d. Commitment. Commitment is devotion to the Corps and your fellow Marines. Without commitment, unit cohesion will break down. Marines must know that their safety and well-being rests upon others. Likewise, Marines must provide the same safety and well-being for themselves and others. Only through this type of commitment will a unit successfully accomplish it mission.

2. **LEADERSHIP CHALLENGES PECULIAR TO MOUNTAIN OPERATIONS**
Although most leadership challenges in a temperate environment are the same in cold weather, some problems will arise which must be quickly corrected.

a. Cocoon-like Existence. Many men, when bundled up in successive layers of clothing and with their head covered by a hood, tend to withdraw within themselves and to assume a "cocoon-like existence". When so clothed, the individual's hearing and field of vision are greatly restricted and he tends to become oblivious to his surroundings. His mental faculties become sluggish and although he looks, he does not see. Leaders must recognize and overcome these symptoms. Additionally, the leader needs to watch for the growth of lethargy within himself and must be alert to prevent it. He must always appear alert to his men and prevent them from sinking into a state of cocoon-like existence.

(1) If your Marines withdraw into a shell or become moody and depressed, get them involved in conversations with each other.

(2) Don't accept an excuse for not carrying out an order. Cold weather training all too often becomes a camping trip. Leadership must challenge their Marines to train as they would fight.

(3) If Marines still display a "Cocoon-like" existence, have them engage in physical activity.

b. Individual and Group Hibernation. This problem is similar in manifestation of withdrawal from the environment. It is generally recognized by a tendency of individuals to seek the comfort of sleeping bags, and by the group remaining in tents or other shelter at the neglect of their duties. In extreme cases, guard and security measures may be jeopardized. Many times, it is the leadership that violates this, thus destroying the unit's trust and confidence.

(1) The leader must ensure that all personnel remain alert and active. Ridged insistence upon proper execution of all military duties and the prompt and proper performance of the many group "chores" is essential.

(2) Be alert for individuals who will place their own physical comfort ahead of their assigned duties. Remind them that their mission as Marines is to fight, and to do so successfully requires that weapons and equipment be maintained in working order.

c. <u>Personal Contact and Communications</u>. It is essential that each individual and group be kept informed of what is happening. Due to the deadening of the senses typically encountered in cold weather, a man left alone may quickly become oblivious to his surroundings, lose his sense of direction, his concern for his unit, and in extreme cases, for himself. He may become like a sheep and merely follows along, not knowing or caring whether his unit is advancing or withdrawing. Each commander must take strong measures to ensure that small unit leaders keep their subordinates informed. This is particularly true of the company commanders keeping their platoon commanders informed, of platoon commanders informing their squad leaders, and the squad leaders informing their men. General information is of value, but the greatest importance must be placed on matters of immediate concern and interest to the individual. The chain of command must be rigidly followed and leaders must see that no man is left uninformed as to his immediate surrounding and situation.

(1) If your Marines find it hard to remember things they have been taught, show patience and review orders, drills and SOP's. Keep their minds busy.

(2) Tempers normally flare up during this type of training, so expect and be prepared to deal with it when it comes. Maintain your sense of humor, lead by example, and don't let unanticipated problems get the best of you.

d. <u>Time/Distance Factors</u>. Mountain operations doctrine recommends that tactical commanders be given every opportunity to exploit local situations and take the initiative when the opportunity is presented. Because of the increased amount of time involved in a movement and the additional time required to accomplish even simple tasks, deviation from tactical plans is difficult. Tactical plans are developed after a thorough reconnaissance and detailed estimate of the situation. Sufficient flexibility is allowed each subordinate leader to use his initiative and ingenuity in accomplishing his mission. Time lags are compensated for by timely issuance of warning orders, by anticipating charges in the tactical situation, and the early issuance of frag orders. Recognition of time/distance factors is the key to successful tactical operations in cold weather mountainous regions.

(1) <u>Time-Distance Formula (TDF)</u>. This formula is designed to be a guideline and should not be considered as the exact amount of time required for your movement. Furthermore, this formula is for use in ideal conditions. The TDF is made for troop on foot in the summertime or troop on skis in the wintertime. If on foot in deep snow, multiply the total time by 2.

(a) 3 km/ph + 1 hour for every 300 meters ascent; and/or +1 hour for every 800 meters descent.

e. Conservation of Energy. Two environments must be overcome in mountainous regions; one created by the enemy and the second created by the climate and terrain. The climatic environment must not be permitted to sap the energy of the unit to a point where it can no longer cope with the enemy. The leader must be in superior physical condition. He cannot expend the additional energy required by his concern for his men and still have the necessary energy to lead and direct his unit in combat. He must remember that there are seldom any tired units, just ***TIRED COMMANDERS!***

(1) IF the unit can effectively fight upon reaching the objective, then it has properly conserved energy.

"It has been repeatedly demonstrated that at temperatures lower than -10F, all other problems lose significance in the personal battle for SURVIVAL"

3. **Survival ASPECTS OF LEADERSHIP**. When dealing with leadership challenges in a survival situation, the foremost weapon a leader must employ is his vigilance: a leader's attention should be focused on ensuring all Marines of the unit are contributing to the overall success of the situation.

a. Cohesion. As a leader, you must ensure that all members of the team are working towards the survivability of the unit. You can not allow individuals or small groups to formulate their own goals or plan of action.

b. Self-Worth. A Marine without self-worth is a Marine who does not value living. Leadership is a critical; factor in building self-worth. Tasks must be found for each Marine in which best suits their situation while attempting to receive positive results. (i.e., A man with a broken leg can monitor the fire, A man with a broken arm can still procure water for the unit). This will make each and every Marine feel useful and not a burden to the other members, regardless of their individual situation.

c. Natural Reactions to Stress. Leadership must quickly identify natural reactions to signs of stress his Marines may be displaying (i.e., Fear, Anxiety, Guilt, Depression). Failure to recognize these signs early will result in injuries, illness, or death which will reduce the unit's combat effectiveness. Corrective action must be taken immediately4

d. Will to Survive. The will to survive is a "mind-set" that must be instilled and reinforced within all Marines. Without the "will to survive", Marines will not succeed. The following tools can aid to develop this "mind-set".

(1) The Code of Conduct.

(2) Pledge of Allegiance.

(3) Faith in America.

(4) Patriotic Songs

(5) Spiritual Faith.

4. **CONCLUSION**. Paramount to survival is preparation and training that will foster trust and confidence in a unit's capability to beat the elements and the enemy. Poorly trained units will not possess the "Will to Survive" as they lack the fundamental skills to overcome the survival situation. Individual confidence is built through challenging and realistic training that teaches a Marine how to survive and how to effectively employ cold weather equipment.

"Spirit of confidence comes form training and tradition... each individual Marine, because of the fighting tradition of the Corps and the toughness of training, is confident of his own ability and that of his buddies... This confidence in themselves and one anther very often spells the difference between victory and SURVIVAL and defeat and annihilation."
FMFM 1-0 "Leading Marines"

REFERENCE:

1. ALMAR 439196, Core Values
2. MCDP 1-0, Leading Marines
3. FMFM 7-21, Tactical Fundamentals for Cold Weather Warfighting

UNITED STATES MARINE CORPS
Mountain Warfare Training Center
Bridgeport, California 93517-5001

MSVX.02.06
2/6/05

STUDENT HANDOUT

SIGNALING AND RECOVERY

TERMINAL LEARNING OBJECTIVE

(1) In a survival situation, and given a survival kit, conduct a survival recovery, in accordance with the references. **(MSVX.02.06)**

(2) In a survival situation, and given a survival kit, conduct survival signaling, in accordance with the references. **(MSVX.02.18)**

ENABLING LEARNING OBJECTIVES

(1) Without the aid of reference, execute a recovery, in accordance with the references. **(MSVX.02.06a)**

(2) Without the aid of references, describe in writing the audio international distress Signal, in accordance with the references. **(MSVX.02.18a)**

(3) Without the aid of references, describe in writing the visual international distress signal, in accordance with the references. **(MSVX.02.18b)**

(4) Without the aid of references, construct an improvised signaling device, in accordance with the references. **(MSVX.02.18c)**

OUTLINE

1. **SIGNALING DEVICES**. The equipment listed below are items that may be on your body or inside an aircraft. Generally, these items are used as signaling devices while on the move. They must be accessible for use at a moment's notice. Additionally, in a summer mountainous environment, Marines may experience areas that are snow covered and must be familiar with the effects that snow will have on specific signaling devices.

    a. Pyrotechnics. Pyrotechnics include star clusters and smoke grenades. When using smoke grenades in snow pack, some form of floatation must be used. Without floatation, the smoke grenade will sink into the snow pack and the snow will absorb all smoke. Rocket parachute flares and hand flares have been sighted as far away as

35 miles, with an average of 10 miles. Pyrotechnic flares are effective at night, but during daylight their effectiveness is reduced by 90 percent.

b. M-186 Pen Flare. The M-186 Pen Flare is a signaling device carried in the vest of crew chiefs and pilots. **Remember to cock the gun prior to screwing in the flare.**

c. Strobe Light. A strobe light is generally carried in the flight vests of all crew chiefs and pilots. It can be used at night for signaling. Care must be taken at night, because a pilot using goggles may not be able to distinguish a flashing strobe from hostile fire. Therefore, an I.R. cap should be used when possible.

d. Flashlight. By using flashlights, a Morse code message can be sent. An SOS distress call consists of sending three dots, three dashes, and three dots. Keep repeating this signal.

e. Whistle. The whistle is used in conjunction with the audio international distress signal. It is used to communicate with forces on the ground.

f. AN/PRC-90 & AN/PRC-112. The AN/ PRC 90 survival radio is a part of the aviator's survival vest. The AN/PRC-112 will eventually replace the AN/PRC-90 . Both radios can transmit either tone (beacon) or voice. Frequency for both is **282.8 for voice**, and **243.0 for beacon**. Both of these frequencies are on the UHF Band.

g. Day/Night Flare. The day/night flare is a good peacetime survival signal. The flare is for night signaling while the smoke is for day. The older version flare is identified by a red cap with three nubbins while the new generation has three rings around the body for identification during darkness. The flare burns for approximately 20 second while the smoke burns for approximately 60 seconds.

NOTE: Once one end is used up, douse in water to cool and save the other end for future use.

h. Signal Mirror. A mirror or any shiny object can be used as a signaling device. It can be used as many times as needed. Mirror signals have been detected as far away as 45 miles and from as high as 16,000`, although the average detection distance is 5 miles. It can be concentrated in one area, making it secure from enemy observation. A mirror is the best signaling device for a survivor, but it is only as effective as its user. Learn how to use one now, before you find yourself in a survival situation.

(1) Military signal mirrors have instructions on the back showing how to use it. It should be kept covered to prevent accidental flashing that may be seen by the enemy.

(2) Any shiny metallic object can be substituted for a signal mirror.

(3) Haze, ground fog, or a mirage may make it hard for a pilot to spot signals from a flashing object. So, if possible, get to the highest point in your area when flashing. If you can't determine the aircraft's location, flash your signal in the direction of the aircraft noise.

AIMING THE SIGNAL MIRROR

2. METHODS OF COMMUNICATION

a. Audio. Signaling by means of sound may be good, but it does have some limitations:

(1) It has limited range unless you use a device that will significantly project the sound.

(2) It may be hard to pinpoint one's location due to echoes or wind.

(3) International Distress Signal. **(MSVX.02.18a)** The survivor will make six blasts in one minute, returned by three blasts in one minute by the rescuer.

b. Visual. Visual signals are generally better than audio signals. They will pinpoint your location and can been seen at greater distances under good weather conditions.

(1) The visual international distress symbol is recognized by a series of three evenly spaced improvised signaling devices. **(MSVX.02.18b)**

3. **IMPROVISED SIGNALING DEVICES. (MSVX.2.6c).** Improvised signaling devices are generally static in nature. They must be placed in a position to be seen by rescuers. They are made from any resources available, whether natural or man-made.

a. Smoke Generator. The smoke generator is an excellent improvised signaling device. It gives the survivor the flexibility to signal in either day or night conditions. This type of signal has been sighted as far away as 12 miles, with an average distance of 8 miles. Smoke signals are most effective in calm wind conditions or open terrain, but effectiveness is reduced with wind speeds above 10 knots. Build them as soon as time and the situation permits, and protect them until needed.

(1) Construct your fire in a natural clearing or along the edge of streams (or make a clearing). Signal fires under dense foliage will not be seen from the air.

(2) Find two logs, 6 - 10 inches in diameter, and approximately five feet long. Place the two logs parallel to each other with 3 – 4 feet spacing.

(3) Gather enough sticks, approximately two inches in diameter and four feet long, to lay across the first two logs. This serves as a platform for the fire.

(4) Gather enough completely dry branches to build a pyramid fire. The pyramid fire should be 4 feet by 4 feet by 2 feet high.

(5) Place your tinder under the platform.

(6) Gather enough pine bough to lay on top of the pyramid fire. Ensure that you leave a small opening at the bottom, allowing access to the tinder. This will allow you to light the tinder without removing the pine bough.

(7) To light, ignite the tinder through the opening at the bottom. If available, construct a torch to speed up the lighting process, especially for multiple fires.

(8) To create a smoke effect during the day light hours, place the pine bough on the ignited fire.

(9) Placing a smoke grenade or colored flare under the platform will change the color of the smoke generated. Remember, you want the fire to draw in the colored smoke which will create a smoke color that contrasts with the back ground will increase the chances of success.

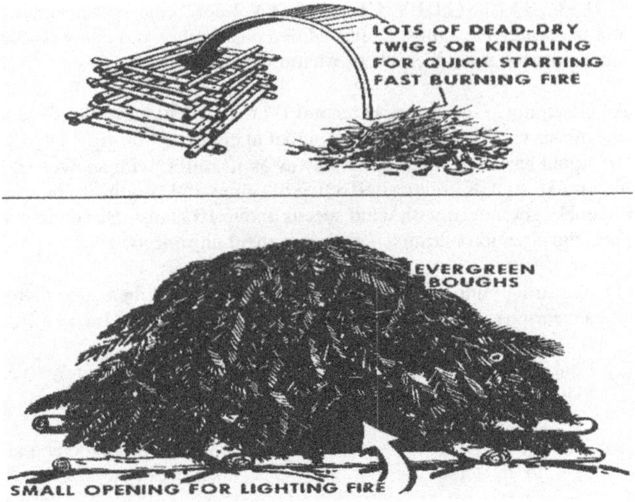

SMOKE GENERATOR

b. Arrangement or alteration of natural materials. Such things as twigs or branches, can be tramped into letters or symbols in the snow and filled in with contrasting materials. To attract more attention, ground signals should be arranged in big geometric patterns.

(1) International symbols. The following symbols are internationally known.

Number	Message	Code symbol
1	REQUIRE ASSISTANCE	V
2	REQUIRE MEDICAL ASSISTANCE	X
3	NO OR NEGATIVE	N
4	YES OR AFFIRMATIVE	Y
5	PROCEED IN THIS DIRECTION	↑

INTERNATIONAL SYMBOLS

(1) Shadows. If no other means are available, you may have to construct mounds that will use the sun to cast shadows. These mounds should be constructed in one of the International Distress Patterns. To be effective, these shadow signals must be oriented to the sun to produce the best shadows. In areas close to the equator, a North—South line gives a shadow anytime except noon. Areas further north or south of the equator require the use of East—West line or some point of the compass in between to give the best result.

(2) Size. The letters should be large as possible for a pilot or crew to spot. Use the diagram below to incorporate the size to ratio for all letter symbols.

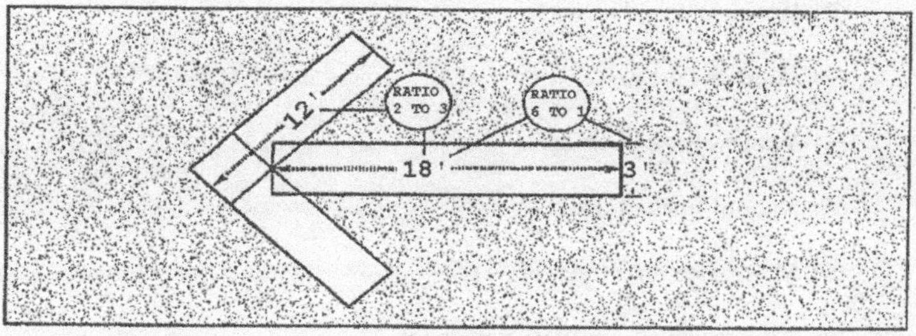

SIZE AND RATIO

(3) Contrast. When constructing letter symbols, contrast the letter from the surrounding vegetation and terrain. Ideally, bring material from another location to build the letter. This could be clothing, air panels, space blanket, etc.

(a) On snow, pile pine bough or use a sea dye marker from an LPP (Life preserver, personal). Fluorescent sea dye markers have been sighted as far away as 10 miles, although the average detection distance is 3 miles.

4. **AIR TO GROUND COMMUNICATIONS**. Air to ground communications can be accomplished by standard aircraft acknowledgments.

a. Aircraft will indicate that ground signals have been seen and understood by:

(1) Rocking wings from side to side. This can be done during the day or in bright moonlight.

b. Aircraft will indicate that ground signals have been seen but **not** understood by:

(1) Making a complete, clockwise circle during the day or in bright moonlight.

5. **RECOVERY**. Marines trapped behind enemy lines in future conflicts may not experience quick recovery. Marines may have to move to a place that minimizes risk to the recovery force. No matter what signaling device a Marine uses, he must take responsibility for maximizing the recovery force's safety.

a. Placement Considerations. Improvised signaling devices, in a hostile situation, should not be placed near the following areas due to the possibility of compromise:

(1) Obstacles and barriers.

(2) Roads and trails.

(3) Inhabited areas.

(4) Waterways and bridges.

(5) Natural lines of drift.

(6) Man-made structures.

(7) All civilian and military personnel.

b. Tactical Consideration. The following tactical considerations should be adhered to prior to employing any improvised signaling device.

(1) Use the signals in a manner that will not jeopardize the safety of the recovery force or you.

(2) Locate a position that affords observation of the signaling device, avenues of approach and provides concealed avenues of escape (if detected by enemy forces). Position should be located relatively close to extract site in order to minimize "time spent on ground" by the recovery force.

(3) Maintain continuous security through visual scanning and listening while signaling devices are employed. If weapon systems are available, signaling devices should be covered by fire and/or observation.

(4) If enemy movement is detected in the area, attempt to recover the signaling device, if possible.

(5) Employ improvised signaling devices only during the prescribed times, if briefed in the mission order.

c. Recovery Devices. **(MSVX.2.6d)** In mountainous terrain, a helicopter landing may be impossible due to ground slope, snow pack, or vegetation. The survivor must be familiar with recovery devices that may be aboard the aircraft.

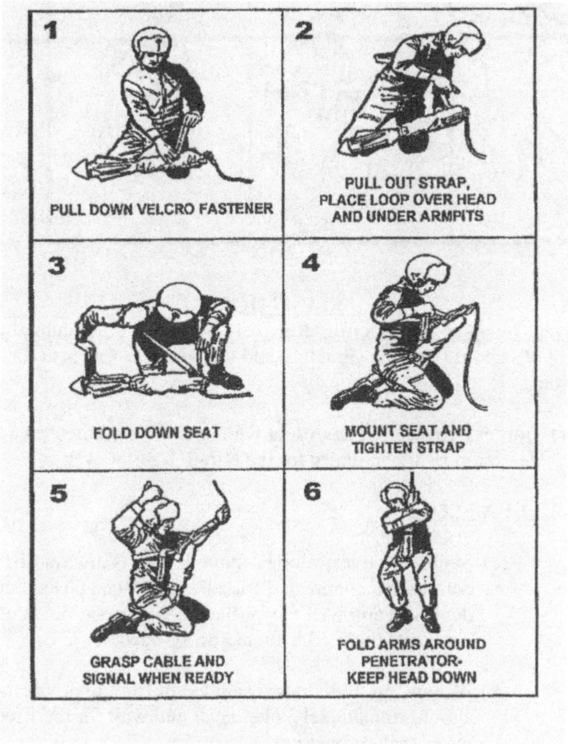

JUNGLE PENETRATOR

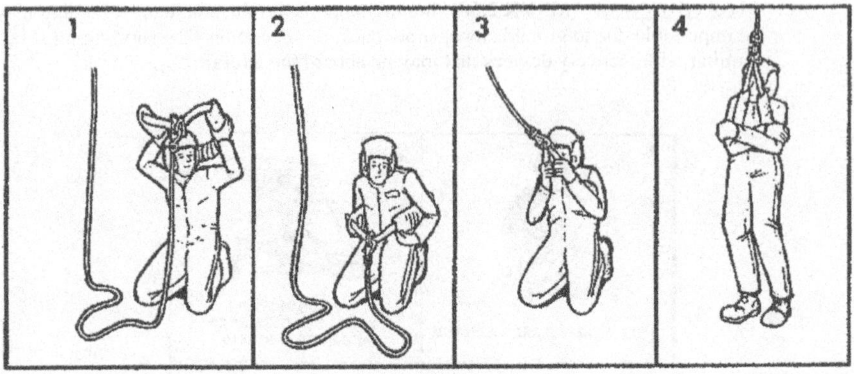

SLING HOIST

d. Recovery by other than aircraft. Recovery by means other than aircraft may occur. Unit SOP's should include signaling and link-up with forces at the following locations:

(1) Border Crossings. The evader who crosses into a neutral country is subject to detention by that country for the duration of the war.

(2) FEBA/FLOT.

(a) Static. Recovery along a static FEBA is always difficult. Under these conditions, enemy and friendly forces can be expected to be densely deployed and well camouflaged, with good fields of fire. Attempts to penetrate the FEBA should be avoided.

(b) Advancing. Individuals isolated in front of advancing friendly units should immediately take cover and wait for the friendly units to overrun their position.

(c) Retreating. Individuals between opposing forces should immediately take cover and wait for enemy units to pass over their position. After most enemy units have moved on, evaders should try to link up with other isolated friendly elements and return to friendly forces.

(3) Link-up with friendly patrols. Unit authentication numbers and/or locally developed codes may assist the evader to safely make contact in or around the FEBA and when approached by friendly forces.

REFERENCE:

1. FM 21-76, Survival, 1992

2. JP 3-50.1, National SAR Manual Volume II, 1991

3. JP 3-50.3, Evasion and Recovery, 1996

UNITED STATES MARINE CORPS
Mountain Warfare Training Center
Bridgeport, California 93517-5001

MSVX.02.07
2/6/05

STUDENT HANDOUT

SURVIVAL NAVIGATION

TERMINAL LEARNING OBJECTIVE In a survival situation, and given a survival kit, employ field expedient navigational aids, in accordance with the references. **(MSVX.02.07)**

ENABLING LEARNING OBJECTIVES

(1) Without the aid of references, list in writing the considerations for travel, in accordance with the references. **(MSVX.02.07a)**

(2) Without the aid of references, describe in writing the seasonal relationship of the sun and its movement during the equinox and solstice, in accordance with the references. **(MSVX.02.07b)**

(3) Without the aid of references, and given a circular navigational chart and operating latitude, determine the bearing of the sun at sunrise and sunset, in accordance with the references. **(MSVX.02.07c)**

(4) Without the aid of references, construct a pocket navigator, in accordance with the references. **(MSVX.02.07d)**

(5) Without the aid of references, describe in writing the methods for locating the North Star, in accordance with the references. **(MSVX.02.07e)**

(6) Without the aid of references, navigate with a pocket or a coal burned bowl, in accordance with the references. **(MSVX.02.07f)**

OUTLINE

1. **CONSIDERATIONS FOR STAYING OR TRAVELLING**. **(MSVX.02.07a)**

a. Stay with the aircraft or vehicle if possible. More than likely somebody knows where it was going. It is also a ready-made shelter.

b. Leave only when:

(1) Certain of present location; have known destination and the ability to get there.
(2) Water, food, shelter, and/or help can be reached.

(3) Convinced that rescue is not coming.

c. If the decision is to travel, the following must also be considered:

(1) Which direction to travel and why.

(2) What plan is to be followed.

(3) What equipment should be taken.

(4) How to mark the trail.

(5) Predicted weather.

d. If the tactical situation permits leave the following information at the departure point:

(1) Departure time.

(2) Destination.

(3) Route of travel/direction.

(4) Personal condition.

(5) Available supplies.

2. DAYTIME SURVIVAL NAVIGATION

a. Sun Movement. It is generally taken for granted that the sun rises in the east and sets in the west. This rule of thumb, however, is quite misleading. In fact, depending on an observer's latitude and the season, the sun could rise and set up to 50 degrees off of true east and west.

b. The following diagram and terms are essential to understanding how the sun and stars can help to determine direction:

Position of the Sun at Equinox and Solstice

(1) Summer/Winter Solstice: (21 June/21 December) Two times during the year when the sun has no apparent northward or southward motion.

(2) Vernal/Autumnal Equinox: (20 March/23 September) Two times during the year when the sun crosses the celestial equator and the length of day and night are approximately equal.

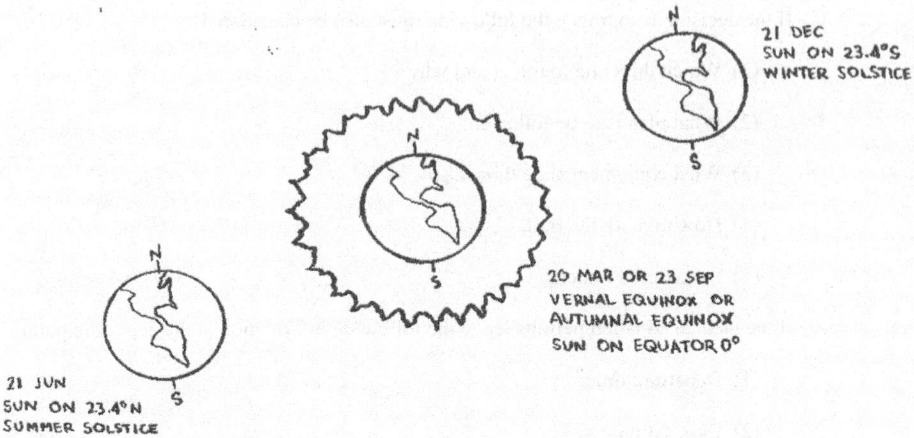

c. Sun's Movement. **(MSVX.02.07b)** As reflected in the diagram above, the earth continuously moves in a cycle from solstice to equinox; throughout each day, however, the sun appears to travels in a uniform arc across the sky from sunrise to sunset. Exactly half-way along its daily journey, the sun will be directly south of an observer (or north if the observer is in the Southern Hemisphere). This rule may not apply to observers in the tropics (between 23.5 degrees north and south latitude) or in the polar regions (60 degrees latitude). It is at this point that shadows will appear their shortest. The time at which this occurs is referred to as "local apparent noon."

d. Local Apparent Noon. Whenever using any type of shadow casting device to determine direction, "local apparent noon" (or the sun's highest point during the day) must be known. Local apparent noon can be determined by the following methods.

(1) Knowing sunrise and sunset from mission orders, i.e., sunrise 0630 and sunset 1930. Take the total amount of daylight (13 hours), divide by 2 (6 hours 30 minutes), and add to sunrise (0630 plus 6 hours 30 minutes). Based on this example, local apparent noon would be 1300.

(2) Using the string method. The string method is used to find two equidistant marks before and after estimated local apparent noon. The center point between these two marks represents local apparent noon.

e. Sun's Bearing. **(MSVX.02.07c)** With an understanding of the sun's daily movement, as well as its seasonal paths, a technique is derived that will determine the true bearing of the sun at sunrise and sunset. With the aid of a circular navigational chart, we can accurately navigate based on the sun's true bearing:

(1) Determine the sun's maximum amplitude at your operating latitude using the top portion of the chart.

(2) Scale the center baseline of the chart where 0 appears as the middle number; write in the maximum amplitude at the extreme north and south ends of the baseline.

(3) Continue to scale the baseline; you should divide the baseline into 6 to 10 tick marks that represent equal divisions of the maximum amplitude.

(4) From today's date along the circumference, draw a straight line down until it intersects the baseline.

(5) The number this line intersects is today's solar amplitude. If the number is left of 0, it is a "north" amplitude; if the number is right of 0, it is a "south" amplitude. Use the formula at the bottom of the chart to determine the sun's bearing at sunrise or sun set.

LATITUDE (N OR S)	5	10	15	20	25	30	35	40	45	50	55	60
MAXIMUM AMPLITUDE	24°	24°	24°	25°	26°	27°	29°	31°	34°	38°	44°	53°

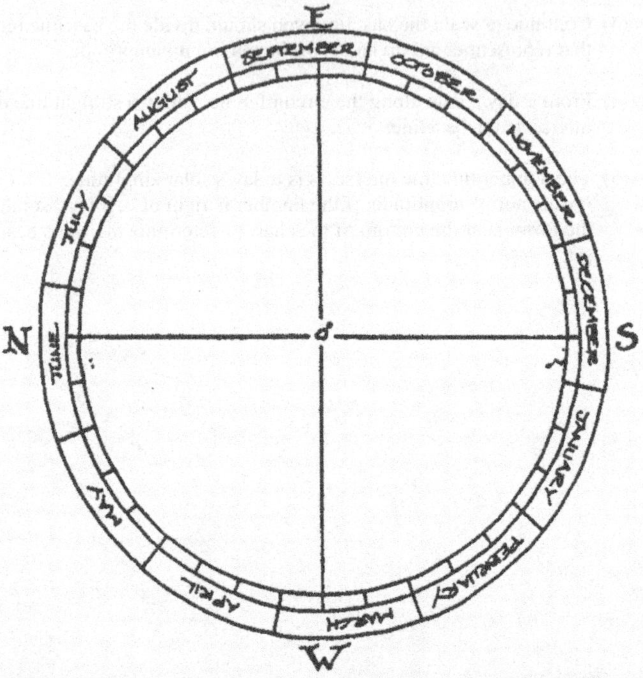

NORTH AMPLITUDES	SOUTH AMPLITUDES
SUNRISE BEARING = 90° − AMPLITUDE	SUNRISE BEARING = 90° + AMPLITUDE
SUNSET BEARING = 270° + AMPLITUDE	SUNSET BEARING = 270° − AMPLITUDE

Circular Navigational Chart

f. Shadow Stick Construction. This technique will achieve a cardinal direction within 10 degrees of accuracy if done within two hours of local apparent noon. Once again, this technique may be impractical near the polar regions as shadows tend to be very long; similarly, in the tropics shadows are generally very small.

(1) Get a straight, 3-6 foot stick free of branches and pointed at the ends and 3-5 small markers: i.e., sticks, rocks, or nails.

(2) Place the stick upright in the ground and mark the shadow tip with a marker.

(3) Wait 10-15 minutes and mark shadow tip again with a marker.

(4) Repeat this until all of the markers are used.

SHADOW STICK METHOD

(5) The markers will form a West—East line.

(6) Put your left foot on the first marker and your right foot on the last marker, you will then be facing north.

3. **POCKET NAVIGATOR**. **(WSV.2.7d)** The only material required is a small piece of paper or other flat-surface material upon which to draw the trace of shadow tips and a 1 to 2 inch pin, nail, twig, wooden matchstick, or other such device to serve as a shadow-casting rod.

   a. Set this tiny rod upright on your flat piece of material so that the sun will cause it to cast a shadow. Mark the position where the base of the rod sits so it can be returned to the same spot for later readings. Secure the material so that it will not move and mark the position of the material with string, pebbles, or twigs, so that if you have to move the paper you can put it back exactly as it was. Now, mark the tip of the rod's shadow.

   b. As the sun moves, the shadow-tip moves. Make repeated shadow-tip markings every 15 minutes. As you make the marks of the shadow tip, ensure that you write down the times of the points.

   c. At the end of the day, connect the shadow-tip markings. The result will normally be a curved line. The closer to the vernal or autumnal equinoxes (March 21 and September 23) the less pronounced the curvature will be. If it is not convenient or the tactical situation does not permit to take a full day's shadow-tip markings, your observation can be continued on subsequent days by orienting the pocket navigator on the ground so that the shadow-tip is aligned with a previously plotted point.

   d. The markings made at the sun's highest point during the day, or solar noon, is the north—south line. The direction of north should be indicated with an arrow on the navigator as soon as it is determined. This north-south line is drawn from the base of the rod to the mark made at solar noon. This line is the shortest line that can be drawn from the base of the pin to the shadow-tip curve.

   e. To use your pocket navigator, hold it so that with the shadow-tip is aligned with a plotted point at the specified point. i.e.; if it is now 0900 the shadow-tip must be aligned with that point. This will ensure that your pocket navigator is level. The drawn arrow is now oriented to true north, from which you can orient yourself to any desired direction of travel.

   f. The pocket navigator will work all day and will not be out of date for approximately one week.

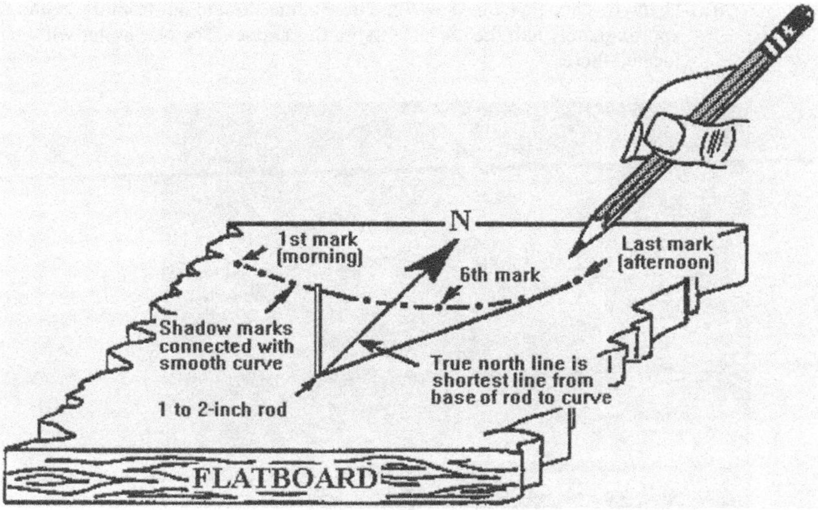

POCKET NAVIGATOR

4. NIGHTTIME SURVIVAL NAVIGATION

a. Mark North. To aid you in navigating at night, it is beneficial to watch where the sun goes down. If you're going to start moving after dark mark the northerly direction.

b. Locating the North Star. There are two methods used in locating the North Star. **(MSVX.02.07e)**

(1) Using the Big Dipper (*Ursa Major*). The best indictors are the two "dippers ". The North Star is the last star in the handle of the little dipper, which is not the easiest constellation to find. However, the Big Dipper is one of the most prominent constellations in the Northern Hemisphere. The two lowest stars of the Big Dipper's cup act as pointers to the North Star. If you line up these two stars, they make a straight line that runs directly to the North Star. The distance to the North Star along this line is 5 times that between the two pointer stars.

(2) Using Cassiopeia (*Big M or W*). Draw a line straight out from the center star, approximately half the distance to the Big Dipper. The North Star will be located there.

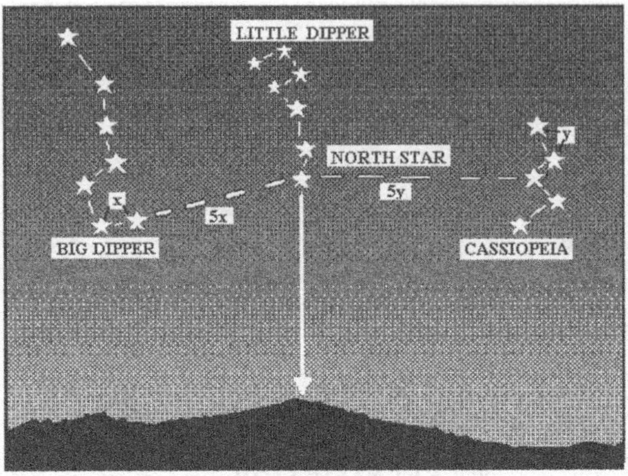

LOCATING THE NORTH STAR

NOTE: Because the Big Dipper and Cassiopeia rotate around the North Star, they will not always appear in the same position in the sky. In the higher latitudes, the North Star is less effective for the purpose of orienting because it appears higher in the sky. At the center of the Arctic circle, it would be directly overhead, and all directions lead South.

c. Southern Cross. In the Southern Hemisphere, Polaris is not visible. There, the Southern Cross is the most distinctive constellation. An imaginary line through the long axis of the Southern Cross, or True Cross, points towards a dark spot devoid of stars approximately three degrees offset from the South Pole. The True Cross should not be confused with the larger cross nearby know as the False Cross, which is less bright, more widely spaced, and has five stars. The True Cross can be confirmed by two closely spaced, very bright stars that trail behind the cross piece. These two stars are often easier to pick out than the cross itself. Look for them. Two of the stars in the True Cross are among the brightest stars in the heavens; they are the stars on the southern and eastern arms. The stars on the northern and western arms are not as conspicuous, but are bright.

Note: The imaginary point depicted in the picture is the dark spot devoid of stars.

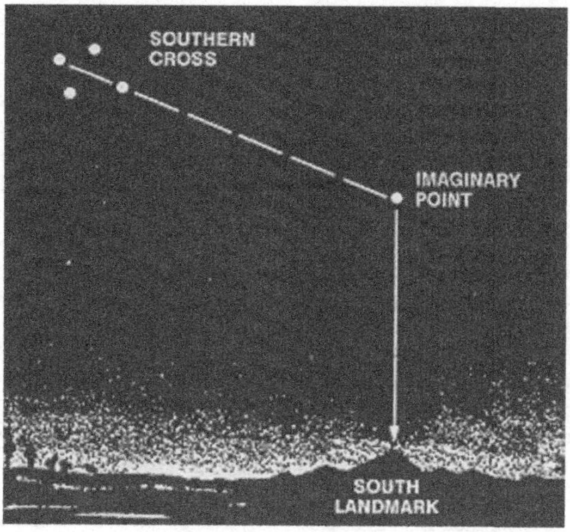

SOUTHERN CROSS

d. Moon Navigator. Like the sun, the moon rises in the east and sets in the west. Use the same method of the shadow stick as you did during the day.

5. **IMPROVISED COMPASSES**. There are three improvised techniques to construct a compass.

a. Synthetic technique. The required items are a piece of synthetic material, (i.e., parachute cloth), and a small piece of iron or steel that is long, thin, and light. Aluminum or yellow metals won't work (only things that rust will do). A pin or needle is perfect, but a straightened paper clip, piece of steel baling wire, or barbed wire could also work.

(1) Stroke the needle repeatedly in one direction against the synthetic material. Ensure that you lift the material a few inches up into the air at the end of each stroke, returning to the beginning of the needle before descending for another

stroke in the same direction. Do this approximately 30 strokes. This will magnetize the needle.

(2) Float the metal on still water using balled up paper, wood chip, or leaf. Gather some water in a non-magnetic container or a scooped out recess in the ground, such as a puddle. Do not use a "tin can" which is made of steel. (An aluminum can would be fine.) Place the float on the water, then the metal on it. It will slowly turn to orient itself.

b. Magnet technique. You will achieve the same results by using a magnet. Follow the same steps as you did with the synthetic material. The magnets you are most likely to have available to you are those in a speaker or headphones of a radio.

c. Magnetization through a battery. A power source of 2 volts or more from a battery can be used with a short length of insulated wire to magnetize metal. Coil the wire around a needle. If the wire is non-insulated, wrap the needle with paper or cardboard. Attach the ends to the battery terminals for 5 minutes.

d. Associated problems with improvised compasses. The following are common problems with all improvised compasses.

(1) Soft steel tends to lose its magnetism fairly quickly, so you will have to demagnetize your needle occasionally, though you should not have to do this more than two or three times a day.

(2) Test your compass by disturbing it after it settles. Do this several times. If it returns to the same alignment, you're OK. It will be lined up north and south, **though you will have to determine by other means which end is north. Use the sun, stars, or any other natural signs in the area.**

(3) Remember, this will give magnetic north. In extreme northern lattitudes, the declination angle can be extreme.

6. **NATURAL NAVIGATION.**

a. Find out where the prevailing winds originate.

b. Sun's path in Northern Hemisphere is SE-SW

1. Bend in trees because of prevailing winds.

2. Sapling Coloration: whiter on one side, darker green on the other. The sunny side (south side) will cause the tree to turn whitish which is a natural sunscreen. White will be on the SW to SE side of the tree. Pick one that is in the open, exposed to the elements all day.

3. Hottest side of a slope will enhance growth: thicker vegetation the SW side.

4. Snow melt on one prominent side of the tree: melt/freeze will indicate the south side.

5. Bleach Rock: the sun's rays has a bleaching effect, lighter side will be to the south. Obviously white rocks are just white rocks.

c. Look for more than one sign to confirm your direction.

7. SURVIVAL NAVIGATION TECHNIQUES

a. Navigator.

(1) Employ a navigation method.

(2) Find the cardinal direction.

(3) Pick a steering mark in the desired direction of travel.

b. Maintain a Log. The possibility may arise when you will not have a map of the area. A log will decrease the chance of walking in circles.

(1) Construction.

(a) Use any material available to you i.e., paper, clothing, MRE box, etc.

(b) Draw a field sketch annotating North, prominent terrain features, and friendly/enemy position.

(2) Maintenance.

(a) Annotate distance traveled, elevation gained and lost, and cardinal directions.

(b) Maintain and update field sketch as movement progresses.

(c) Ensure readability of your field sketch. (i.e.; don't clutter the sketch so much that it can't be read.)

c. During Movement Constantly Refer To.

(1) Log.

(2) Steering marks.

d. Actions If You Become Lost.

(1) Immediate action

(a) Orient your sketch. This will probably make your mistake obvious.

(2) Corrective action

(a) Backtrack using steering marks until you have determined the location of your error.

(b) Re-orient your sketch.

(c) Select direction of travel and continue to march.

REFERENCE:

1. FM 21-76, Survival, 1992.

2. AFM 64-5, Survival, 1969.

3. David Seidmond, The Essential Wilderness Navigator, 1995.

UNITED STATES MARINE CORPS
Mountain Warfare Training Center
Bridgeport, California 93517-5001

MSVX.02.08
2/6/05

STUDENT HANDOUT

SURVIVAL TRAPS AND SNARES

TERMINAL LEARNING OBJECTIVE In a survival situation, and given a survival kit, employ a trap or snare, in accordance with the references. **(MSVX.02.08)**

ENABLING LEARNING OBJECTIVES

(1) Without the aid of references, list in writing the general considerations to take game, in accordance with the references. **(MSVX.02.08a)**

(2) Without the aid of references, list in writing the general techniques to take game, in accordance with the references. **(MSVX.02.08b)**

(3) Without the aid of references, list in writing the requirements for snaring, in accordance with the references. **(MSVX.02.08c)**

(4) Without the aid of references, employ a snare, in accordance with the references. **(MSVX.02.08d)**

(5) Without the aid of references, list in writing the types of triggers, in accordance with the references. **(MSVX.02.08e)**

(6) Without the aid of references, employ a trap, in accordance with the references. **(MSVX.02.08f)**

(7) Without the aid of references, employ a noise producing path guard, in accordance with the references. **(MSVX.02.08g)**

OUTLINE

1. **GENERAL CONSIDERATIONS TO TAKE GAME (MSVX.02.08a)**

a. General Considerations. Knowing a few general hints and tips will make the trapping of animals easier and considerably more effective. Most of these considerations relate to the "ambush mentality". Think about each consideration as if you are planning an ambush of an enemy unit. The eight general considerations to take game are:

(1) Know your game. Knowing the habits of the animal you want to trap or snare will help increase your chances. Such things as when and where they move, feed, and water will help you determine where the set can be most effectively placed.

(2) Keep things simple. You don't have time in a survival situation to construct elaborate sets and they do not necessarily do a better job.

(3) Place sets in the right place. Animals will travel and stop in certain locations. That is where to build sets.

(4) Cover up your scent. Animals will avoid a set, which smells threatening or unusual to them (i.e., human scent or P.O.L.'s from equipment or clothing).

(a) Man leaves a scent through the pores of the skin by the sweat glands. Use an odorless contact glove when building a set. It may take up to three days for your scent to dissipate if made without gloves.

(b) Certain boot soles and clothing may leave a scent, generally this can be detected by the human nose. If noticed, attempt to mask the scent with smoke from your fire

(5) Use the right type of set. Certain sets work better than others do for a particular animal.

(6) Use the correct equipment. Using the correct equipment is paramount to success. This includes the weight of the lifting device in proportion to the animal's weight, the cordage or wire to hold the animal's strength, and trigger tension.

(7) Check traps. Check your traps twice daily: morning and evening. Checking your traps less than twice a day can allow your game to escape, rot, or be taken by other predators.

(8) Lure your sets. Lures will add to your chances of success. Certain lures are appropriate at certain times of the year, depending upon the animal desired. Lures fall into four different categories:

(a) Bait Lures.

(b) Call Lures (these are audio devices).

(c) Gland or Territorial Lures.

(d) Curiosity Lures.

2. **GENERAL TECHNIQUES TO TAKE GAME**. **(MSVX.02.08b)** A general technique is the method in which the trap is intended to kill the animal. The acronym "SICK" is useful in learning these techniques.

(1) S- Strangle. This method strangles the animal, such as a snare.

(2) I- Impale. This method pushes a stake through the animal, such as a spiked dead fall.

(3) C- Crush. This method crushes the animal, such as a deadfall for a chipmunk.

(4) K- Knock. This method knocks a larger animal unconscious, such as a deadfall.

3. **SNARE NOMENCLATURE AND IMPROVISED SNARES** A snare in nothing more than a piece of wire, rope, or cord with a loop at one end, which tightens down around animal's neck. Snares are much easier and less time consuming to construct than traps, while producing better results.

(1) Wire. Although snares can be used with rope or cord, they are less effective than wire. Wire should have memory and resist kinking. Aircraft cable type 7x7, in sizes 1/16 to 3/8 inch should be used. This type of wire prevents animal chew out and resists breakage. Remember that you want the smallest diameter cable capable of holding the animal.

(2) Locking device. A locking device is imperative for a snare to work properly. Locking devices secure the snare around the animal's neck. There are several methods available for a locking device.

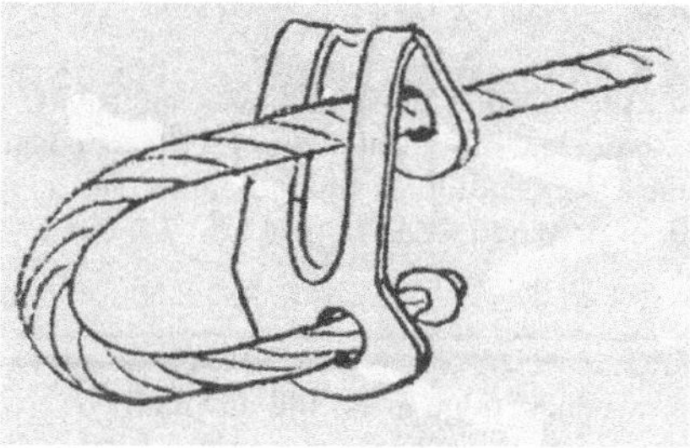

COMMERCIAL

BENT WASHER

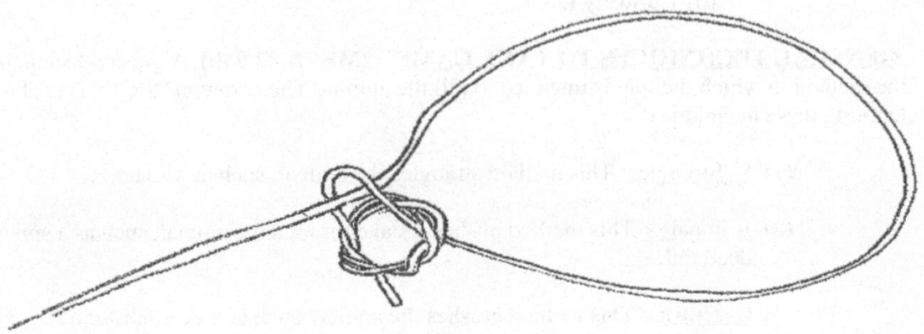

KNOT

4. SNARING. In any environment, snaring is the most effective means to take game. A snare in nothing more than a piece of wire, rope, or cord with a loop at one end which tightens down around an animals' neck. Snares are much easier and less time consuming to construct than traps, while producing better results.

   a. Requirements for snaring **(MSVX.02.08c)** There are three requirements too effectively employ snares. They are location, presentation, and construction.

   b. Location. Location is paramount to success. The following guide can assist you.

      (1) Coyotes, Foxes, and Badger. Along rarely traveled roads, fire lanes, irrigation ditches, fence lines, in saddles, along ridge tops, meadow borders, by carcasses.

      (2) Bobcat and Lynx. At bases of cliffs and large rock faces; on ridges and saddle crossing; along stream bottoms. Cats need security so rarely get far from trees or brush, or escape cover.

      (3) Raccoon, Opossum, Skunk, and Ring-tailed Cats. Stream beds and banks; trails along stream beds, ponds, rivers and other water courses; Raccoons like a combination of water, old mature trees, buildings and junk piles, and a consistent food supply like grain or prepared feed.

      (4) Weasel. Marshy, grassy meadows.

      (5) Marten and Fisher. Along meadow edges, ridge lines, and downfalls.

      (6) Mink. Under bridges; around culverts, tiles and junk in or near streams, rivers and lakes, springs and seeps, Muskrat and Beaver lodges and dams. Mink will stop and investigate nearly every hole or cavity around a streambed.

(7) Beaver and Muskrat. Around the food cache under the ice.
(8) Rabbits. In thick willow stands along runs and trails.

c. Presentation. Presentation is the type of set for the intended animal. Cubbie set works well for bobcat, raccoon, marten, fisher, opossum, and skunk while trail sets work well for coyote, fox, mink, and rabbit.

d. Construction. Construction is the actual building of the set.

(1) A split stick is utilized to support the snare and ensure the snare fires properly. The split stick can be either green or seasoned wood. However, the snare must not slip through the split. The locking device must be next to the split stick. The split stick must be securely placed in the ground. (If you are utilizing a weighted snare the locking device may be in the 12 o'clock position.)

(2) The snare must be anchored or attached to a drag.

(3) The snare must be **loaded** so it will fire quickly.

(4) Loop size. A correctly employed snare will have the snare holding the animal around the neck. Loop size is placed on the snare according to the intended animal. Too large will result in a body or leg catch, resulting in possible chew out or breakage. Too small will enable the animal to force the snare to the side, resulting in a miss. Additionally, the loop must be placed with specific ground clearance. Ideally, the bottom of the loop should hit the intended animal chest high. The snare trigger is that part of the loop which hits the animals chest.

ANIMAL	NOOSE SIZE	GROUND CLEARANCE
SQUIRREL	2 1/2 TO 3 INCHES	1/2 TO 1 1/2 INCHES
RABBIT	4 TO 5 1/2 INCHES	1 1/2 TO 3 INCHES
RACCOON	6 INCHES	3 TO 4 INCHES
FOX	7 TO 10 INCHES	8 TO 10 INCHES
COYOTES	12 TO 14 INCHES	12 INCHES
BOBCAT	9 INCHES	8 INCHES

Note: (1) Noose size is the diameter of the snare loop.
(2) Ground clearance is measured from the bottom of the loop to the ground.

(5) Fencing. The objective of fencing is to have the animal move through the "path of least resistance" or more importantly the snare. Fencing must be subtle and not over done.

(6) Lure. In a survival situation, you will not be able to employ numerous snares. Luring all snares increases your chances of success.

(a) Bait Lures.
-MRE peanut butter, cheese spread, or jelly.
-Dead carcasses.
-Dead rodents.

(b) Gland or Territorial Lures.
-Animal Urine mixed with beaver castors or animal glands.

(c) Curiosity Lures.

-Single feather, bird wing, piece of fur tied and suspended under a tree limb and allowed to freely move with the breeze.

5. **SNARE SETS**. **(MSVX.02.08d)** Although there are numerous ideas to employ snares, here are a few.

TRAIL SET

CUBBIE SET

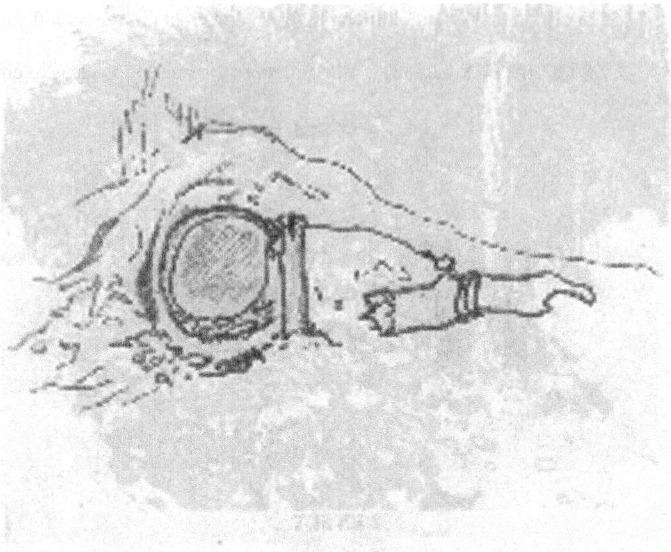

DEN SET

VARIOUS LOG SETS

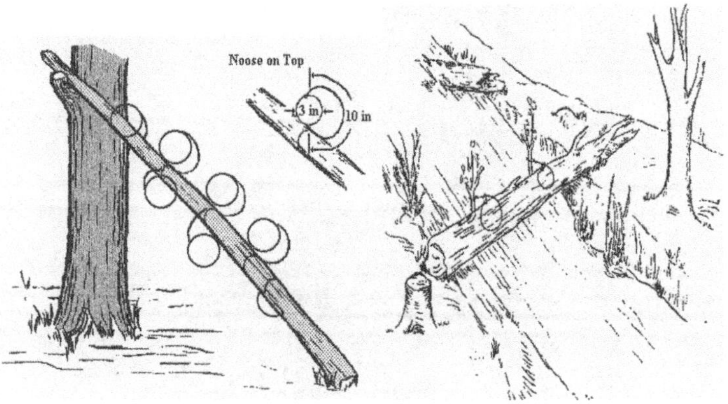

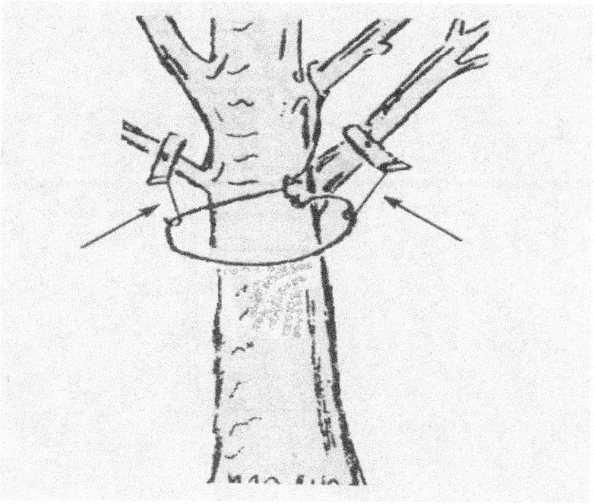

6. **<u>IMPROVISED TRAPS</u>**. Improvised traps are made from a variety of materials. These traps are designed to hold or kill animals by use of some type of action. This action is generally caused by either a weight or spring loaded device.

a. <u>Triggers</u>. **(MSVX.02.08e)** There are three basic triggers used for all traps and pathguards. Depending on the situation, variations of these triggers can be used. The key to all improvised traps is the trigger system.

(1) <u>Modified Puite figure 4</u>. The Puite deadfall requires a knife and piece of cord to construct. It is designed to mangle small rodents. The trigger for the deadfall is the modified Puite figure 4.

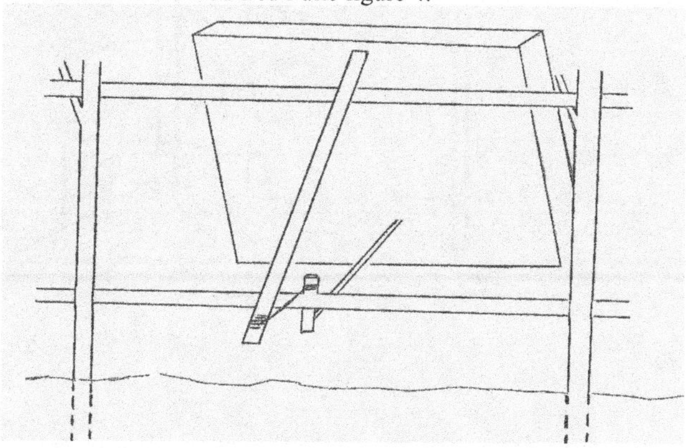

<u>MODIFIED PUITE DEADFALL</u>

(2) <u>Toggle</u>.

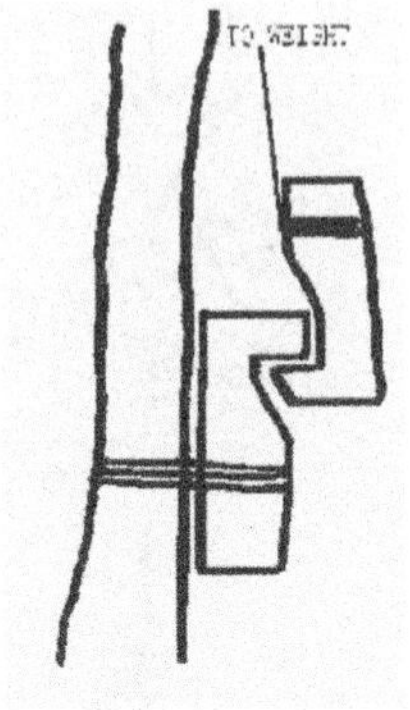

(3) Universal.

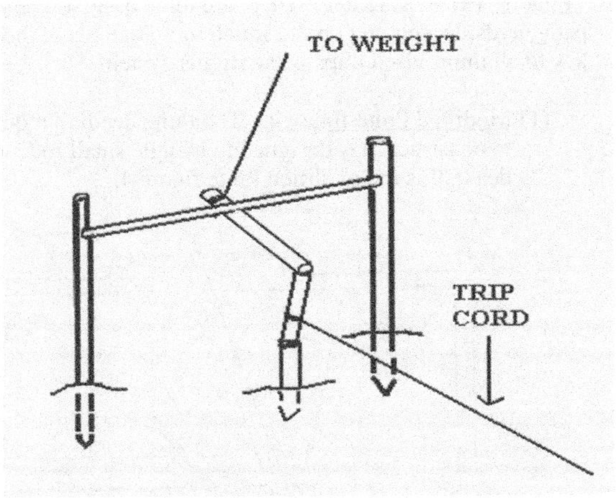

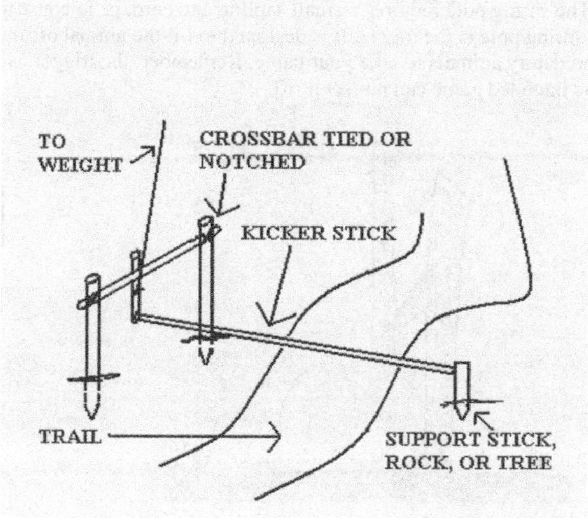
TO WEIGHT
CROSSBAR TIED OR NOTCHED
KICKER STICK
TRAIL
SUPPORT STICK, ROCK, OR TREE

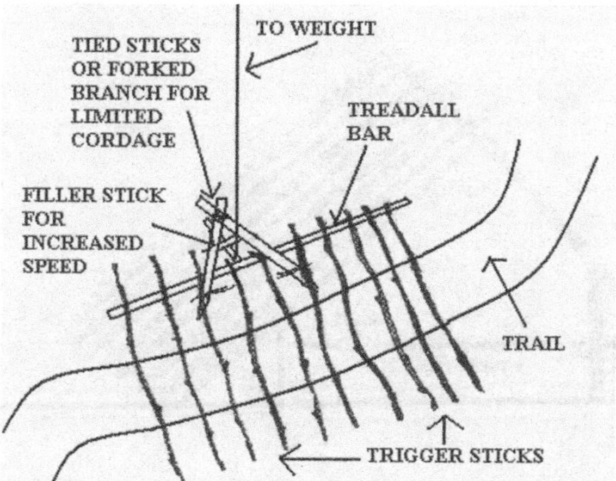
TO WEIGHT
TIED STICKS OR FORKED BRANCH FOR LIMITED CORDAGE
TREADALL BAR
FILLER STICK FOR INCREASED SPEED
TRAIL
TRIGGER STICKS

b. Spring Pole. The spring pole requires a small sapling and cordage to construct. The trigger for the spring pole is the toggle. It is designed to lift the animal off the ground; not allowing predatory animals to take your game. Remember, the trigger can not be so tight that the intended game can not set it off.

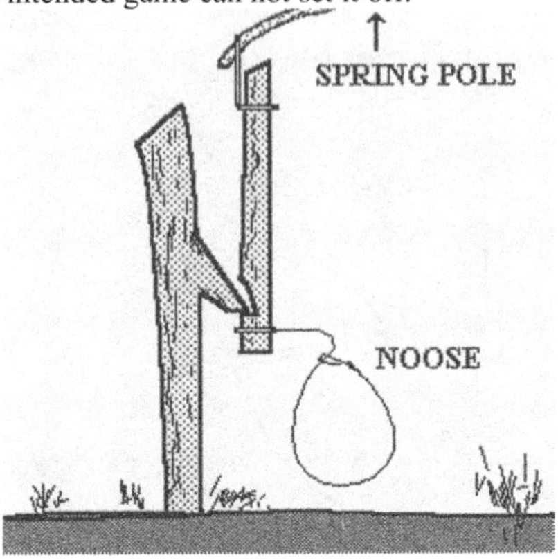

SPRING POLE

c. Box Trap. The box trap requires limited cordage to construct. It is designed to hold live small rodents and birds. The box trap trigger is the Puite figure 4.

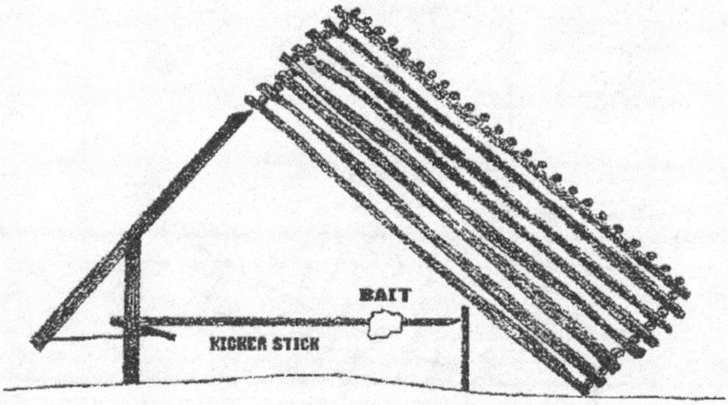

BOX TRAP

d. Baited Treble Hook. Tie a large treble hook onto a tree limb high enough to cause the animal to jump but not so high it cannot reach it. Bait the treble hook.

7. **PATH GUARDS (MSVX.02.08g)** Path guards are designed to protect and provide security for your shelter area against the enemy and predatory animals. They are classified into noise and casualty producing path guards.

a. Noise producing path guards. Noise producing pathguards serve as an alarm for your shelter area. When triggered, it should produce some type of loud noise or visual signal. Although construction can vary, depending on materials available, one example is as follows:

(1) Secure a young sapling to a universal trigger.

(2) At the end of the sapling, tie several pieces of metal to the sapling. Use whatever is available for metal.

(3) Camouflage the metal on the ground.

(4) When triggered, the sapling should swing back and forth, causing the metal to rattle.

b. Casualty producing path guards. Casualty producing path guards, when triggered, should cause death or injury to the enemy or predatory animal. Tips should be poisoned as discussed in Survival Plant Uses class. Triggers for this type of path guard should be the universal trigger.

(1) Log Jerk.

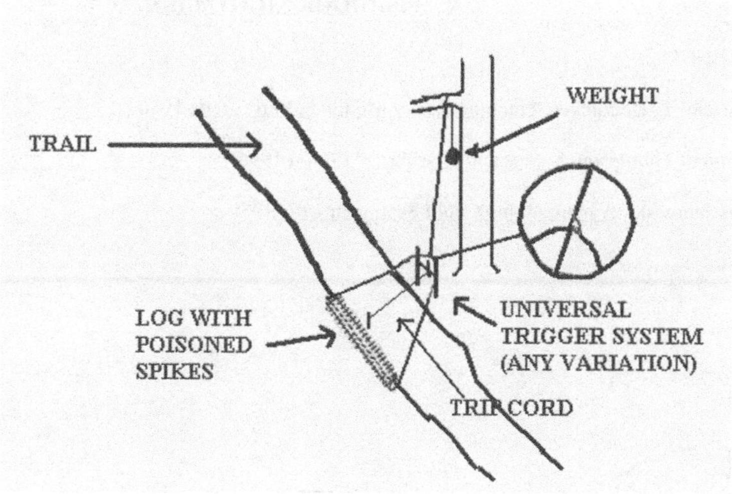

(2) Fish Hook Nightmare

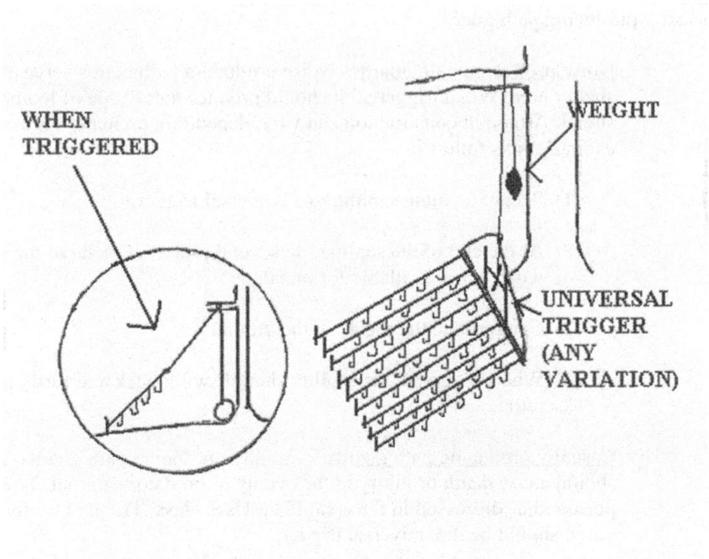

FISHHOOK NIGHTMARE

REFERENCE:

1. Dr. Major L. Boddicker, Trapping Rocky Mountain Furbearers, 1980.

2. Raymond Thompson, Snares and Snaring, 3rd Edtion 1996.

3. Chris Janowski, A manual that Could Save your Life, 1989

UNITED STATES MARINE CORPS
Mountain Warfare Training Center
Bridgeport, California 93517-5001

MSVX.02.09
02/06/05

STUDENT HANDOUT

SURVIVAL USES OF GAME

TERMINAL LEARNING OBJECTIVE In a survival situation, and given a survival kit, and procured game, prepare game for consumption, in accordance with the references. **(MSVX.02.09)**

ENABLING LEARNING OBJECTIVES

(1) Without the aid of references, and given an animal, dress and skin game, in accordance with the references. **(MSVX.02.09a)**

(2) Without the aid of references, and given a green hide, construct a suitable product, in accordance with the references. **(MSV.02.09b)**

(3) Without the aid of references, and given a piece of meat, preserve meat, in accordance with the references. **(MSVX.02.09c)**

(4) Without the aid of references, list in writing the acronym "TOM", in accordance with the references. **(MSVX.02.09d)**

(5) Without the aid of references, list in writing the parts of game that can be used for human consumption, in accordance with the references. **(MSVX.02.09e)**

OUTLINE:

1. **KILLING GAME**.

A. Nose Tap and Heart Stomp. Using a club hit the animal on the nose. This will knock the animal unconscious. Lay the club across the animal's neck. Placing one foot on the club to keep the animal down. Then use the heel of your other boot to give the animal several sharp blows to the chest area . This causes the heart to swell up and the animal bleeds internally.

B. Bleeding. Slicing the animal's throat or piercing the chest cavity accomplishes this method.

C. Bludgeoning. Simply beat the animal until it stops moving.

D. Breaking the Neck. The first step is the same as the nose tap. Laying the club across the neck, pull on the rear legs until a "snap" is heard and release. Once the animal straightens its hind legs, it is dead.

2. **PREPARING GAME** **(MSVX.02.09a)**

A. Dressing. Once the animal is dead, dressing should occur immediately. This allows the chest cavity to cool, thus slowing the decay and bacteria rate. Use of chest cavity propping sticks will aid in this process.

1) Game

a) Using a well-sharpened pocket knife, cut around anus. Be careful not to puncture intestines or kidneys.

b) Cut the hide from the anus towards the chest cavity. This is performed by first placing two fingers under the skin. Next, place the blade of the pocket knife in between your fingers. This prevents rupturing the intestines and contaminating the meat.

c) Reach in and pull out the heart, lungs, and liver, keeping them separate from the guts. These organs are edible. Check the liver for white spots. If white spots appear on the liver, the animal may have tularemia.

2) Birds

a) Pluck feathers while body is warm or the bid can be dipped into hot water.

b) A bird can also be skinned. However, this process removes the birds fat layer and is wasteful in a survival situation.

c) Make incision from vent to tail and draw out intestines.

3) Reptiles/Amphibians

(a) Cut off head well down behind poison sacs.

(b) Cut open skin from anus to neck. Pull out internal organs and discard.

Note: Box turtles, brightly colored frogs, frogs with "X" mark on their backs, and toads should be avoided.

B. Skinning. Although, the hide acts as a protective layer it should be removed as soon as possible. This will allow the meat to cool and develop a glaze. Since blood is a food source it should be collected when possible.

(1) Large Game - Caping

(a) Find the Achilles tendon just above the feet and cut a small hole between the bone and the tendon. Now you can thread a rope, string, etc., through the hole in order to hang the animal upside down from a tree branch or a make shift rack.

(b) Cut completely around the hind legs, just below where the animal is suspended. Then cut towards the anus on the inside of the hind legs.

(c) Pull hide straight down towards the head. The procedure used on the hind legs will be repeated for the fore legs. Continue pulling the hide until it is free of the head. The hide will have to be cut if the animal has antlers.

(2) Small Game - Casing

(a) Small game can be skinned like large game or it can be cased. Casing a hide means to pull the entire skin off the carcass from rear forward, with cuts made only around the feet of the animal and from the back legs to the tail. This method allows the skin to be made into mittens, bags, and other holding devices.

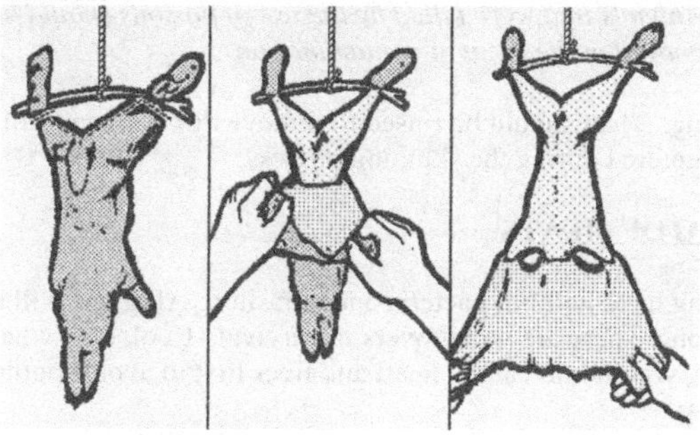

HIDE CASING

(3) Fish

(a) The skin of fish is usually left on.

(4) Birds

(a) The skin of birds should be left on. There is a heavy layer of fatty tissue between skin and meat.

(5) Reptiles/Amphibians

(a) The skin of reptiles is left on.

C. Butchering. Is simply cutting the meat into manageable portions. Smaller animals are generally best left whole.

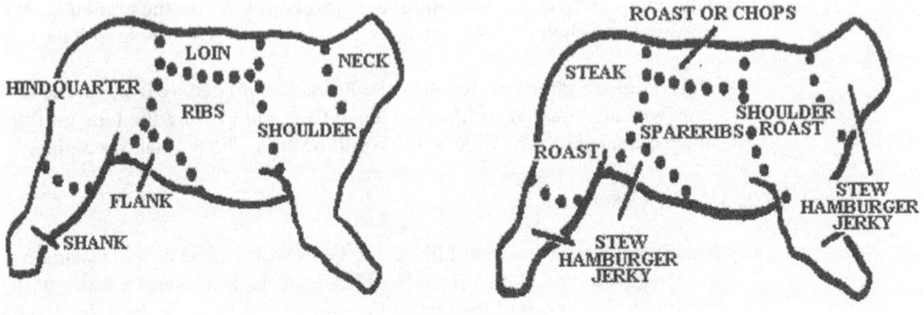

BUTCHERING

Note: *Animals that were killed by the use of poisons should have a 2" cubic size square of meat removed at the point of contamination.*

D. Washing. Meat should be rinsed to remove dirt and especially if any bladder or fecal organs were ruptured during the skinning process.

3. COOKING OF MEATS

A. Cooking meat will kill bacteria and parasites. All game will be cooked until it is thoroughly well done. There are no leftovers in survival. Cook only what you can immediately consume. Ideally, you should eat the heart and liver first to avoid spoiling. There are two methods of cooking.

1) Boiling. This is the best method for cooking. Boiling enables the survivor to consume the animal fat and nutrients, which collect in the broth.

2) **Roasting. This method is wasteful and will not be utilized in a survival situation.**

4. TANNING HIDES (MSVX.02.09b)

A. Fleshing. Fleshing is the actual removal of meat, tissue, and fat from the hide. Fleshing is easier when done as soon as possible, preferably before the hide starts to dry.

1) Soak or wet the hide if dry.

2) Lay the hide on a solid, smooth, round object (i.e., log or canteen).

3) Holding a bayonet, blunt knife, sharp stone or bone tool scrapper at a 10 degree angle away from the body, push the fat and membrane off the leather. Be careful not to make holes in the more tender parts of the belly. For beaver and badger the fat must be cut off the hide.

4) Continue this until all the fat is completely removed.

B. Stretching. A fresh green or soaked hide must be stretched. Stretching is accomplished by either making a frame or using the ground.

1) Frame stretching involves lacing the hide to a frame with cordage and pulling it tight.

2) Ground stretching involves staking the hide tight to the ground.

3) A Frame can also be created by bending a stick back on it's self.

4) The less a hide shrinks and hardens the softer it will be at the end of the process.

C. Hair.

1) Using a sharp stone tool scrapper scrape off the hair. Soaking the hide in water will make this process easier.

D. Braining. The brain acts as a lubricant and provides a temporary water repellant.

1) Soak the hide on the stretcher.

2) Extract the brain from the animal.

3) Mix the brains with water to create a pasty solution.

4) Once the brains are warm and thoroughly mixed rub into the hide. Firmly rub the mixture into hide with your hand on only the hairless side.

E. Graining. This step forces the brains thoroughly into the leather.

1) Sponge on water to further dampen hide.

2) Using a blunt end of a pole, apply pressure over every inch, scrapping and stretching the fibers until most of the water is gone from the skin.

3) Tighten the hide on the stretcher and allow it to dry.

F. Rubbing. The next critical step is the high friction rubbing needed to create a little heat and finish the drying, stretching, and breaking of the grain. Either method can be used.

1) Cut the skin from the frame around the perimeter, leaving only the lacing holes and hair that could not be removed.

2) Use a one half inch rope attached between two trees. Grasp the skin at different points all around its perimeter and pull, pull, pull.

3) If rope is not available, rub the skin by sitting on the ground and hooking the skin over your feet and pulling.

G. Smoking. Smoking the hide will help make the hide water repellant.

1) Add wet or green wood chips to the fire. Sage or willow are good woods. The object is to get the chips to smoke, not burn.

2) It only takes a few minutes to smoke, but be careful to prevent flame from ruining the hide.

H. Animal Hide Uses. Animal hide uses are limited only by the imagination. Listed below are a few ideas:

5. **PRESERVING MEATS (MSV.02.09c)**

A. Botulism. **(MSVX.02.09d)** Botulism is an often-fatal food poisoning caused by improperly preserved meats. Botulism grows in a controlled environment. The acronym "TOM" is useful in defeating botulism. If any one of the three elements is removed from the preserving process, botulism cannot live.

1) Temperature, botulism thrives between 40-140 degrees F.

2) O- Oxygen, botulism needs an airtight environment to live.

3) M- Moisture, botulism needs a moist environment to live.

B. Freezing.

1) Before freezing, cut the meat into pieces of a size that can be used one at a time.

2) Keep it frozen until ready to use. Remember, meat will spoil if thawed and refrozen.

C. Cooling.

1) Place meat in a metal or wooden container with a lid. The container should be ventilated.

2) Set it in water or bury it in damp earth, preferably in a shaded location.

3) Do not throw moldy meat away; cut or scrape off the mold and cook as usual.

D. Jerky. Jerky allows the meat to last a couple of weeks while reducing the weight of meat by dehydrating it. Jerky is made from the meat only.

1) Cut meat into thin strips about 1/4 inch thick. Remove all thick portions of fat.

2) Place meat by a fire to lightly smoke it. You are attempting to develop a thin crust layer on the meat. This serves to deter the bugs and insects. Remember to use hard woods and not conifer type wood. You want to smoke it, not cook the meat.

3) Once the meat has a crust layer, remove the meat and place strips on a hanger for the air to dry it for approximately 24 hours. Once dry, break down fibers by slightly pulling apart the meat and allow it to dry another 24 hours.

4) When it becomes hard and brittle, it is taken down and stored in breathable bags or cloth. It is used in stews, soups, or roasted lightly on coals and eaten.

5) Small animals, fish, and birds are dried whole. After they are skinned, the back is cracked between the legs, a stick is inserted to hold the body cavity open. The animal is lightly smoked and laid out in the sun to dry. When thoroughly dried, they are pounded until the bones are crushed. Another day in the air will dry the marrow and ensure preservation.

E. Pemmican. Pemmican allows meat to last for several months.

1) Dry berries and pound into a paste.

2) Dried jerky is added to the paste.

3) Melted suet (the hard fatty tissues around the kidneys) is mixed with the berries and jerky.

4) Roll the mixture into small balls and place in the cleaned intestines of a large animal.

5) The intestine sack is tied shut, sealed with suet and stored in plastic or leather bags.

6. SPECIFIC PARTS

A. Other than the actual meat on game there are other parts of it that can be eaten. They are the: **(MSVX.02.09e)**

1) Brain.

2) Eyes.

3) Tongue.

4) Liver.

5) Heart.

6) Lungs.

7) Kidneys.

8) Gizzards.

REFERENCE:

1. Paul Auerbach, Wilderness Medicine, 3rd Edition, 1995.
2. Larry Dean Olson, Outdoor Survival Guide, 5th Edition, 1990.

3. B-GA-217-001/PT-001, Down but not Out, Canadian Survival Guide.
4. Wilderness Way, Volume 2, Issue 1.
5. John Wiseman, SAS Survival Guide, 1993.
6. Chris Janowski, A Manual that could save your life, 1989.
7. William R. Davidson, Field Manual of Wildlife Diseases in the Southeastern United States, 2nd Edition, 1997.

UNITED STATES MARINE CORPS
Mountain Warfare Training Center
Bridgeport, California 93517-5001

MSVX.02.10
02/06/05

STUDENT HANDOUT

FIELD EXPEDIENT TOOLS, WEAPONS, AND EQUIPMENT

TERMINAL LEARNING OBJECTIVE In a survival situation, and given a survival kit, construct field expedient implements, in accordance with the references. **(MSVX.02.10)**

ENABLING LEARNING OBJECTIVES

(1) Without the aid of references, list in writing the resources used to construct field expedient tools , in accordance with the references. **(MSVX.02.10a)**

(2) Without the aid of references, list in writing the methods for lashing a handle to a field expedient tool, in accordance with the references. **(MSVX.02.10b)**

(3) Without the aid of references, list in writing the types of clubs, in accordance with the references. **(MSVX.02.10c)**

(4) Without the aid of references, construct a club, in accordance with the references. **(MSVX.02.10d)**

(5) Without the aid of references, construct a bowl, in accordance with the references. **(MSVX.02.10e)**

OUTLINE

1. **RESOURCES** **(MSVX.02.10a)** The materials used to make all field expedient tools, weapons, and equipment will fall into one of the five categories.

   a. Stone.

   b. Bone.

   c. Wood.

   d. Metal.

   e. Other materials.

f. Stone. Stone will make an excellent striking, puncturing or chopping tool, but will not hold a fine edge. Some stones, such as chert, flint, or obsidian can have very fine edges.

(1) Chipping & Flaking. To make a sharp-edge piece of stone, a chipping tool and flaking tool is needed. A chipping tool is a light, blunt-edged tool used to break off small pieces of stone. A flaking tool is a pointed tool used to break off thin, flattened pieces of stone. You can make a chipping tool from wood, bone, or metal, and a flaking tool from bone, antler tines, or soft iron.

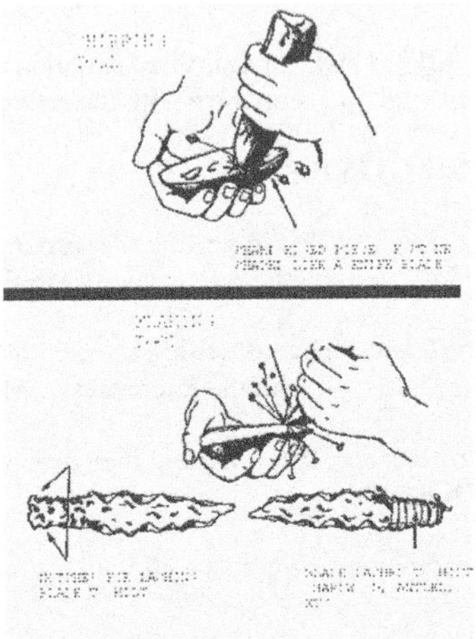

(2) Weapon heads. Certain stones will shatter under pressure when force is delivered upon it. When selecting a stone, test its hardness prior to use.

g. Bone. Bone has many uses. Hooks, shaft tips, scrapers, awls, sockets and handles are just a few ideas.

(1) Raw Bone. Raw bone must be shattered with a heavy object, such as a rock.

(2) Shaping & Sharpening. From the pieces of shattered bone, select a suitable pointed splinter. You can further shape and sharpen this splinter by rubbing it on a rough surfaced rock or metal file (i.e., from your multi-purpose knife).

h. Wood. Wood uses are unlimited. A knife blade can shape the wood into any desired shape.

(1) Types. Wood is classified into two general categories: hard and soft. Hardwood is preferred for all survival uses. To test the wood strength, press your fingernail into the grain of the wood. If a print is visible, the wood is generally soft.

(2) Sharpening. All wood points are sharpened to the side of the shaft. Wood is weakest at the center and will not hold a point.

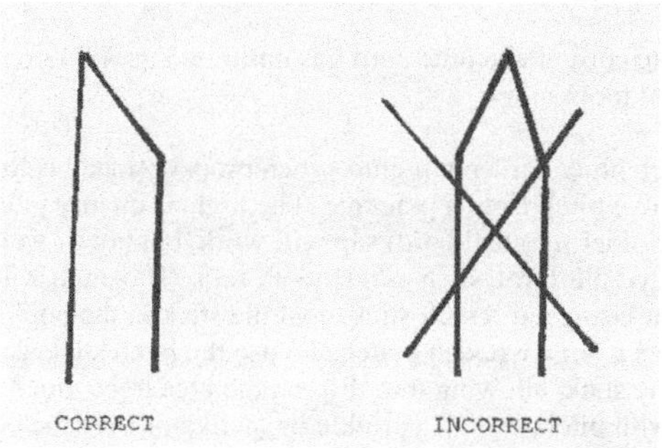

(3) Fire hardening. Wood that is unseasoned or "green" wood should be fire hardened prior to use. To test this wood, gently scrape the bark with your thumbnail. If moisture or a greenish tint appears, it is considered green. Fire harden it by holding the point of the instrument a few inches above a bed of hot coals while slowly rotating it. Gradually the wood will begin to hiss and steam. Fire hardening makes the cells swell and the sap thicken, which makes the wood more resistant to abrasion and concussion. Avoid charring the wood. Fire harden only the tip until light brown.

(4) Coal burning. It is very difficult to carve depressions in wood. A depression in wood can be made by a process called coal-burning. Using a pair of thongs, place a hot coal over the area you want to hollow out, then blow on the embers with a thin, steady stream of air to keep them glowing. If available, use a thin reed or length of hollow bone to direct the stream of air. After the coals have burned down, scrape out the charred wood with a knife or sharp rock. Repeat this process with fresh sets of coals until the depression is at the desired depth.

i. Metal. Metal is the best material to make field expedient edged weapons. When properly designed, metal can fulfill a knife's three uses: puncture, slice or chop, and cut. First, select a suitable piece of metal, one that most resembles the desired end product. Depending on the size and original shape, you can obtain a point and cutting edge by rubbing the metal on a rough surfaced stone or metal file. If the metal is soft enough, you can hammer out one edge while the metal is cold. Use a suitable flat, hard surface as an anvil and a harder object of stone or metal as a hammer to hammer out the edge.

j. Other materials. Other materials are those items that can be found or may be on your body which can be used in the construction of field expedient tools.

(1) Load bearing equipment clips. The sliding retaining clip can be removed and sharpened to a point.

(2) Plastic. Plastic, Plexiglas, and glass from an aircraft can be shaped and sharpened into a point. Plastic can also be melted as a adhesive.

(3) Parachute cord. Parachute cord has unlimited uses for construction of field expedient tools.

(4) Pine pitch glue. Pine pitch glue, when properly made is like an epoxy. Locate and remove pitch from a pine tree. The highest quality pitch to use is fresh sap. The older (dry and hard) sap will work, but not as well. Melt the pitch on an elevated platform, such as a smooth rock. The pitch will run down the platform. Using a 6-8 inch stick, coat the stick in the pool of pitch until it resembles a large wooden match. To use the pitch stick as glue, light the pitch end of the stick, allowing it to drip on the area to be glued. Once sufficiently coated with pitch, sprinkle the activator over the pitch. An activator is finely ground egg shell or fire wood ash.

2. **CLUBS. (MSVX.02.10c)** Clubs are held and not thrown. As a field expedient weapon, the club does not protect you from enemy soldiers. It can, however, extend your area of defense beyond your fingertips. It also serves to increase the force of a blow without injuring yourself. There are two types of clubs: simple and weighted.

a. Simple club. A simple club is a staff or branch. It must be short enough for you to swing easily, but long and strong enough to damage whatever you hit. Its diameter should fit comfortably in the palm, but not be so thin as to break easily upon impact.

b. Weighted club. A weighted club is any simple club with a weight on one end. The weight may be a natural weight, such as a knot on the wood, or something added, such as a stone lashed to the club. If adding a weight to the club, construction is as follows:

(1) Find a stone that has a shape which will allow you to lash it securely to the club. A stone with a slight hourglass shape works well. If a suitably shaped stone cannot be found, you must fashion a groove or channel into the stone by a technique known as pecking. By repeatedly rapping the club stone with a smaller hard stone, you can get the desired shape.

(2) Find a piece of wood that is the right length. Hardwood is the best, if available. The length should feel comfortable in relation to the weight of the stone.

(3) **(MSVX.02.10b)** Lash the stone to the handle. There are two techniques for attaching the stone to the handle: forked and wrapped.

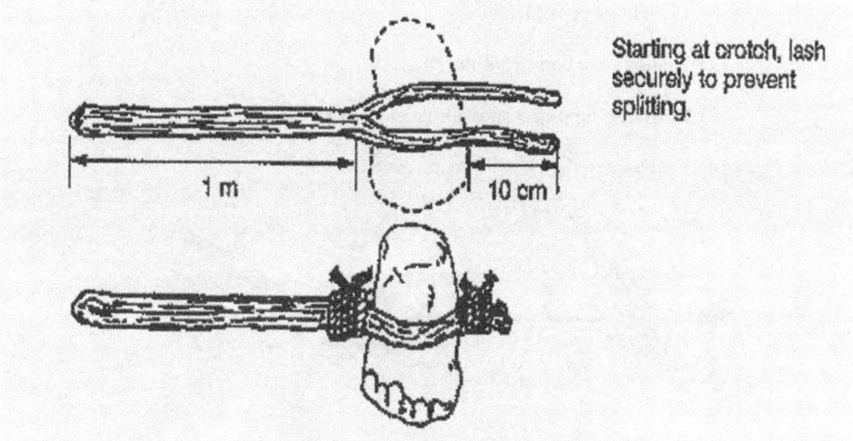

FORKED-BRANCH TECHNIQUE

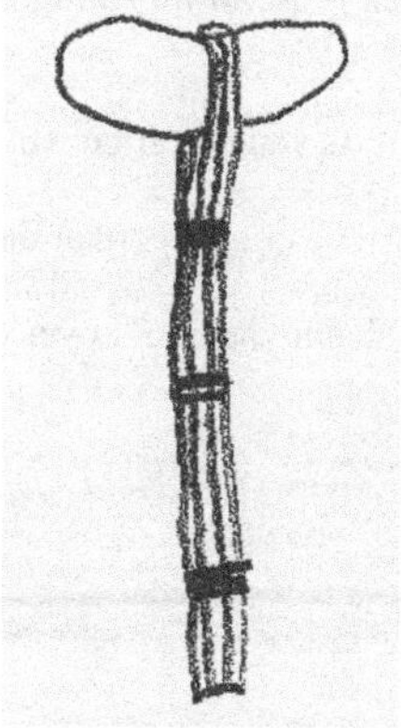

WRAPPED

3. **SURVIVAL STICKS**. There are four types of survival sticks which are useful in a survival situation.

a. Digging stick. Finding an edible root is fairly easy, but most roots grow deep, and digging them out can be difficult unless one is prepared with a few good techniques. Skillfully applied, a simple device called the digging stick saves time and energy that is otherwise expended scrapping and grubbing with flat stones and fingers, which could lead to infection.

(1) Find a hardwood stick that is three feet long, one inch in diameter, and is straight as possible.

(2) Remove the bark from the stick.

(3) Form the tip of the stick into a chisel shape.

(4) Fire harden the chisel if using green wood.

DIGGING STICKS

b. Noose stick. A noose stick is useful for strangling and controlling improperly snared animals that are still alive.

(1) Find a pole as long as you can effectively handle.

(2) Attach a noose of wire or stiff cord at the small end.

(3) To catch an animal, slip the noose over the neck and pull it tight.

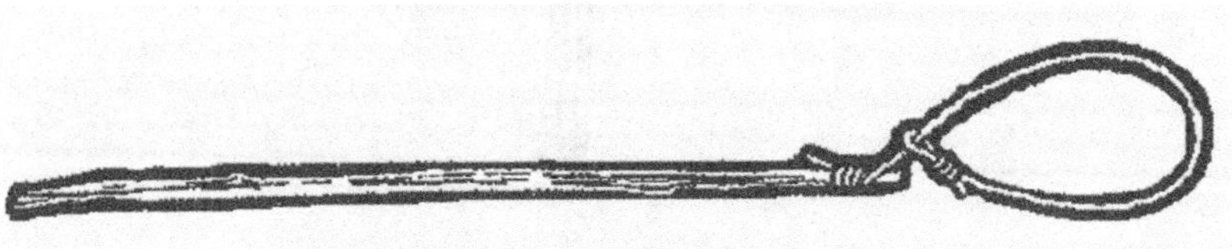

NOOSE STICK

c. Sling shot. A Y-shaped stick can easily be made into a sling shot. A sling shot is an extremely effective and accurate weapon.

(1) Locate a hardwood, Y-shaped piece of stick.

(2) From your survival kit, attach the sling shot rubber and pouch.

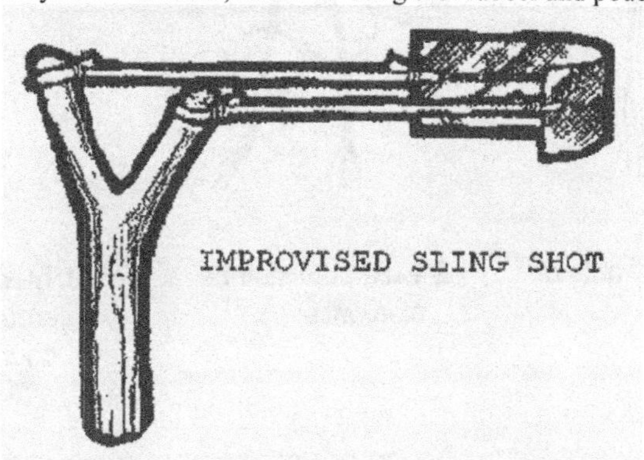

d. Throwing stick. One of the simplest weapons for survival is the throwing stick. As a tool, the throwing stick can be used to knock dead branches out of a tree that would normally be too high to reach. The dead branches can then be used as firewood.

(1) Find a stick straight as possible, 2.5-3 feet long, and 1.5-2 inches in diameter.

(2) Remove the bark from the stick.

(3) Taper each end of the stick.

(4) Fire harden the entire stick if using green wood.

(5) There are two methods of employing the throwing stick. When in forested area, the best method is to use an overhand throwing motion. In an open area, you can increase the killing radius by using a sidearm throwing motion.

4. **Cordage**. Before making cordage, there are a few simple tests that can be done to determine the material's suitability. First, pull on a length of the material to test for strength. Next, twist it between your fingers and roll the fibers together. If it withstands this handling and does not snap apart, tie an overhand knot with the fibers and gently tighten. if the knot does not break, the material is usable.

a. Suitable cordage can be made from Iris leaves, Yucca, or Stinging Nettle stalks.

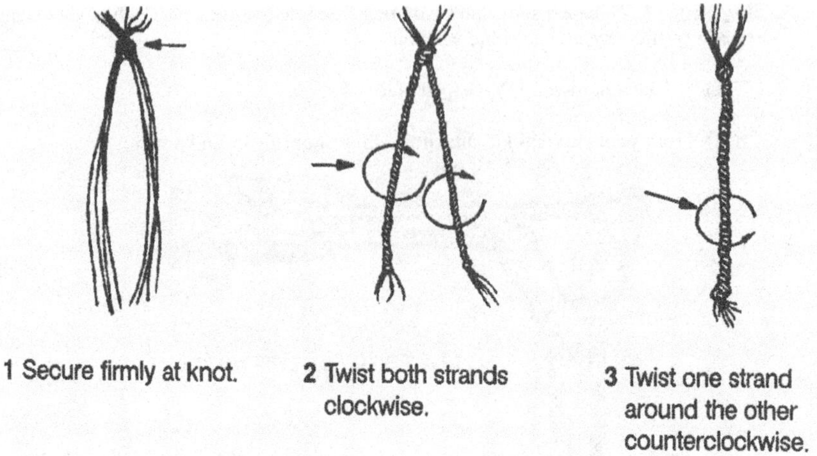

5. **EXPEDIENT PACKS**. The horseshoe pack is simple to make, use, and relatively comfortable to carry over one shoulder.

HORSESHOE PACK

a. Lay available square-shaped material, such as a poncho or tarp flat on the ground.

b. Lay items on one edge of the material. Place those items frequently used (i.e., canteens) on the outside. Pad the hard items.

c. Roll the material (with the items) towards the opposite edge and tie both ends securely.

d. Tie extra lines along the length of the bundle.

e. Fold bundle in half and secure a long piece of rope to the apex of the fold.

f. Attach pack to your body.

6. **UTENSILS**. Utensils are used for cooking, eating, and storing food.

a. Bowl or Container. Bowls and containers can serve to carry and store food. They can be made from bone and wood. To make them out of wood:

(1) Locate or split a piece of wood.

(2) Coal burn to the desired depth.

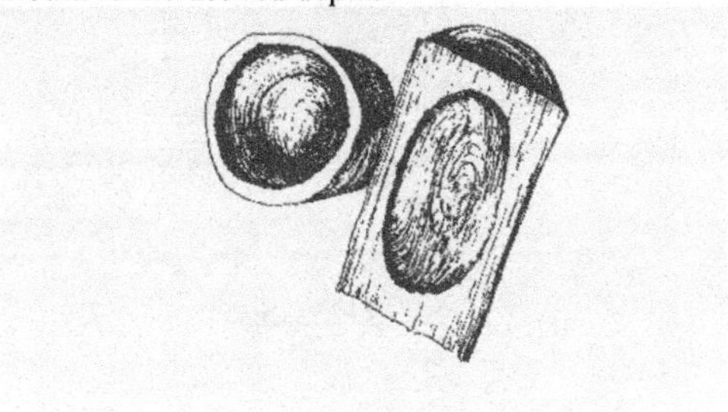

COAL BURNED BOWLS

b. Spork. A spork is a useful tool to eat with. With a knife, carve a piece of wood into the desired shape.

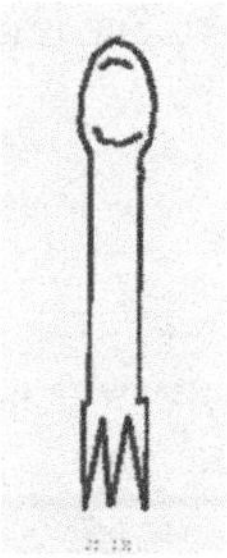

SPORK

c. Tongs. Thongs aid to move hot items, such as coal embers.

(1) Cut a piece of green sapling.

(2) Split the sapling in half and shave off the bark. Flatten both ends of each section.

(3) Fire harden each half.

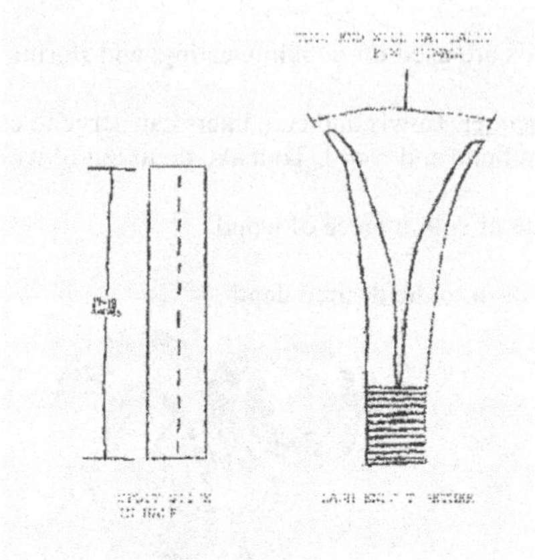

TONGS

Reference:

1. FM 21-76, Survival, 1992.

2. Chris Janowski, A Manual that could save your life, 1989.

3. Tom Brown, Field Guide to Wilderness Survival, 1983.

UNITED STATES MARINE CORPS
Mountain Warfare Training Center
Bridgeport, California 93517-5001

MSVX.02.11
04/19/00

STUDENT OUTLINE

FORAGING PLANTS AND INSECTS FOR SURVIVAL USES

TERMINAL LEARNING OBJECTIVE

(1) In a survival situation, and given a survival kit, identify and subsist on plant resources, in accordance with the references. **(MSVX.02.11)**

(2) In a survival situation, and given a survival kit, identify and subsist on insect resources, in accordance with the references. **(MSVX.02.12)**

ENABLING LEARNING OBJECTIVES

(1) Without the aid of references, list in writing the plants to be avoided, in accordance with the references. **(MSVX.02.11a)**

(2) Without the aid of references, prepare an edible plant, in accordance with the references. **(MSVX.02.11b)**

(3) Without the aid of references, prepare a plant for medicinal purposes, in accordance with the references. **(MSVX.02.11c)**

(4) Without the aid of references, list in writing the plants that can be used to poison animals, in accordance with the references. **(MSVX.02.11d)**

(5) Without the aid of references, list in writing the six insects to be avoided, in accordance with the references. **(MSVX.02.12a)**

(6) With the aid of references, prepare an edible insect, in accordance with the references. **(MSVX.02.12b)**

OUTLINE

1. **GENERAL CONSIDERATIONS**. There are very few regions throughout the world without some type of edible vegetation. Plants contain vitamins, minerals, protein, carbohydrates, and dietary fiber. Some plants also contain fats. The following are general considerations:

a. Do not assume that because birds or animals have eaten a plant, it is edible by humans.

b. Poor plant recognition skills will seriously limit your ability to survive.

c. Plant dormancy and snowfall make foraging plants difficult during the winter months.

d. Plants generally poison by:

(1) Ingestion. When a person eats a part of a poisonous plant.

(2) Contact. When a person makes contact with a poisonous plant that causes any type of skin irritation or dermatitis.

(3) Absorption. When a person absorbs the poison through the skin, which can interrupt a bodily function.

(4) Inhalation. Poisoning can occur through the inhalation of smoke that contains poisonous plant residue.

e. Plant properties can change throughout the growing season. Plants can be edible during certain periods while poisonous in others.

2. **PLANT IDENTIFICATION**. Proficiency in plant identification is complex and requires diligent study. You identify plants, other than by memorizing particular varieties through familiarity, by using such factors as leaf shape and margin, leaf arrangements, and root structure.

a. Leaf Margins. The basic leaf margins are toothed, lobed, and toothless or smooth.

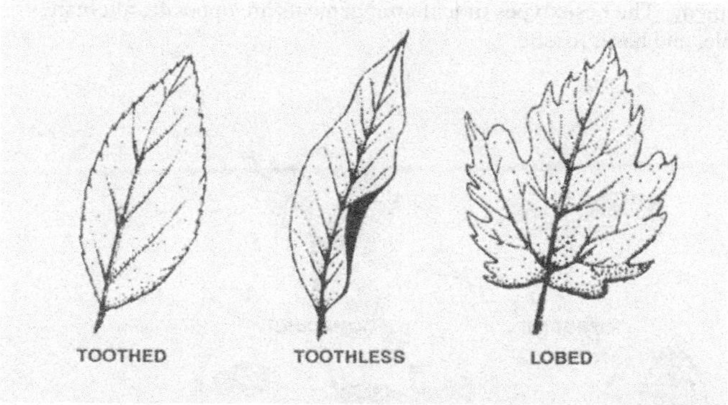

b. Leaf Shape. These leaves may be lance-shaped, elliptical, egg-shaped, oblong, wedge-shaped, triangular, long-pointed, or top-shaped.

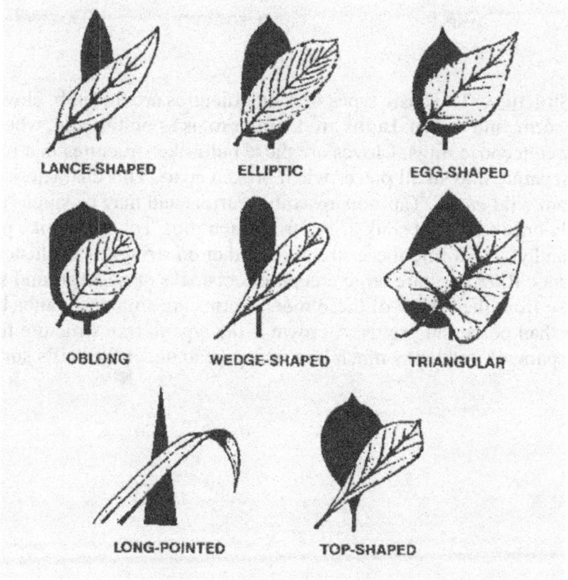

c. Leaf Arrangement. The basic types of leaf arrangements are opposite, alternate, compound, simple, and basal rosette.

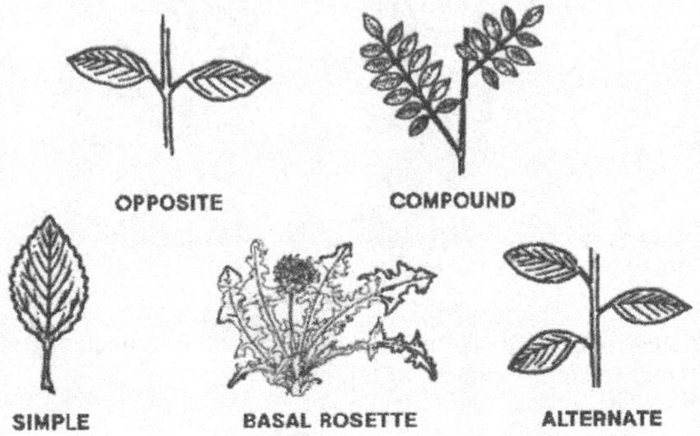

d. Root Structure. The basic types of root structures are the bulb, clove, taproot, tuber, rhizome, corm, and crown. Bulbs are familiar to us as onions and, when sliced in half, will show concentric rings. Cloves are those bulb-like structures that remind us of garlic and will separate into small pieces when broken apart. This characteristic separates wild onions from wild garlic. Taproots resemble carrots and may be single-rooted or branched, but usually only one plant stalk arises from each root. Tubers are like potatoes and daylilies and you will find these structures either on strings or in clusters underneath the parent plants. Rhizomes are large creeping rootstocks or underground stems and many plants arise from the "eyes" of these roots. Corms are similar to bulbs but are solid when cut rather than possessing rings. A crown is the type of root structure found on plants such as asparagus and looks much like a mop-head under the soil's surface.

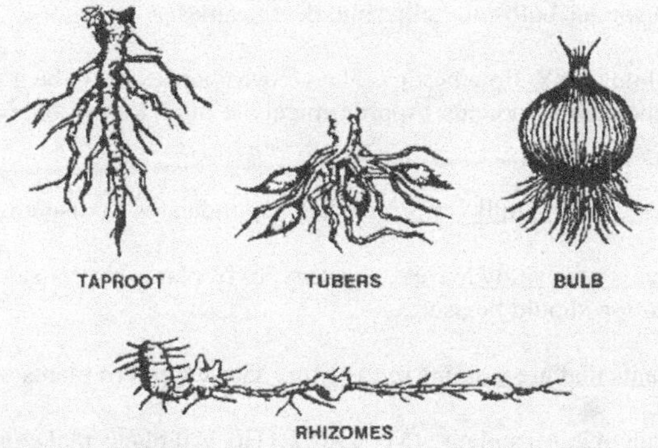

3. **DETERMINING EDIBILITY**. The thought of having a diet consisting only of plant food is often distressing. This is not the case if the survival episode is entered into with the confidence and intelligence based on knowledge or experience. If a Marine knows what to look for, can identify it, and know how to prepare it properly for eating, there is no reason why he can't find sustenance.

   a. Types of plants to avoid. **(MSVX.02.11a)** Experts estimate there are about 300,000 classified plants. There are two considerations that must be kept in mind when procuring plant food. The first consideration is that the plant be edible, and preferably, palatable. Next, it must be fairly abundant in the areas in which it is found. If it includes an inedible or poisonous variety in its family, the edible plant must be distinguishable to the average eye from the poisonous one. Usually a plant is selected because one special part is edible, such as the stalk, the fruit, or the nut. When selecting an unknown plant as a possible food source, apply the following general rules:

      (1) Mushrooms and Fungi. These should not be selected because they have toxic peptides, a protein-base poison which has no taste.

      (2) Plants with umbrella-shaped flowers. These plants are to be completely avoided, although carrots, celery, dill, and parsley are members of this family. One of the most poisonous plants, poison water hemlock, is also a member of this family.

      (3) Beans, Peas, and Seeds in pods. All of the legume family should be avoided (beans and peas). They absorb minerals from the soil and cause problems. The most common mineral absorbed is selenium. Selenium is what has given locoweed its fame.

      (4) 3-leafed and Whorled-leafed growth patterns. These leaf patterns are members of the *Lupinas* genus and other poisonous plants.

(5) All bulbs. As a general rule, all bulbs should be avoided. Examples of poisonous bulbs are tulips and death camas.

(6) White and Yellow berries. The colored berries are to be avoided as they are almost all poisonous. Approximately ½ of all known red berries are poisonous.

(7) Plants with a milky sap. A milky sap indicates a poisonous plant.

(8) Plants with shiny leaves. These types of plants are considered poisonous and caution should be used.

(9) Plants that are irritants to the skin. These types of plants include poison ivy.

b. Preparing an unknown plant. **(MSVX.02.11b)** All plants that **cannot** be positively identified will be prepared properly prior to testing for consumption. Do not prepare plants that are described by types of plants to avoid. Many harmful toxins contained in plants can be destroyed by heat, or are water soluble, though some toxins remain exceptions.

(1) Place one part (leaves, flowers, stems, or roots) in a canteen cup.

(2) Fill the canteen cup with water and boil.

(3) After the water has boiled, remove the canteen cup from the heat source. Strain the water out, leaving the plant inside the canteen cup. Cooking and discarding the water can lessen or remove the amount of toxins that may be contained in the plant. These boiling periods should last at least 5 minutes each.

(4) Fill the canteen cup with water and repeat the process.

(5) After straining the water out a second time, the plant may be tested.

c. Plant testing procedure. The US Military has developed a plant testing procedure, which may not always guarantee the safety of plants for human consumption. This procedure should only be used in a long-term survival situation as a last resort.

(1) Ensure that you haven't eaten any food for at least 8 hours. Prepare one canteen cup of charcoal for emergency consumption.

(2) Select a plant that grows in sufficient quantity in the local area. Separate the part of the plant you wish to test: root, stem, leaf, or flower. Certain parts of plants are poisonous while the other parts may be edible.

(3) Rub a portion of the plant you have selected on your inner forearm. Wait 15 minutes and look for any swelling, rash, or irritation. What you are testing for is contact poisoning.

(4) Prepare the plant using the unknown plant procedure as described above.

(5) Touch a small portion (a pinch) of prepared plant to the outer surface of your lip to test for burning or itching. If after 3 minutes there is no reaction on your lip, place a pinch of the prepared plant food on your tongue and hold for 15 minutes. **Do not swallow**.

(6) After holding on your tongue for 15 minute, if there is no reaction, thoroughly chew and hold it in your mouth for another 15 minutes. **Do not swallow.** If unpleasant effects occur (burning, bitter, or nauseating taste), remove the plant from the mouth at once and discard it as a food source. If no unpleasant effects occur, swallow the plant material and wait 8 hours.

(7) If after 8 hours no unpleasant effects have occurred (nausea, cramps, diarrhea), eat a ¼ cup of plant prepared the same way and wait 8 hours.

(8) In no unpleasant effects have occurred at the end of this 8 hour period, the plant may be considered edible if prepared in the same manner.

(9) If at any time symptoms appear (i.e., nausea, cramps, or dizziness), attempt to induce vomiting and consume prepared charcoal.

(10)Completely document and sketch the plant in a log book to refer to for future use. This will aid in future procurement of this plant. If plant properties have changed, you will have to repeat the plant testing procedure.

4. **EDIBLE PLANTS**. There are many recognizable plants located throughout the world. Remember, eating large portions of a single plant food on an empty stomach may cause diarrhea, nausea, or cramps. Two good examples of this are such familiar foods as green apples and wild onions; therefore, eat them in moderation.

a. General Considerations. The following are general considerations which can be applied throughout the world.

(1) Select plants resembling those cultivated by people.

(2) Single fruits on a stem are generally considered safe to eat.

(3) Blue or black berries are generally safe for consumption.

(4) Aggregated fruits and berries are always edible (for example, thimble berry, raspberry, salmonberry, and blackberry).

(5) Plants growing in the water or moist soil are often the most palatable.

(6) All cone-bearing trees produce seeds in their cones which are edible. The cambium layer, pitch, and pine needles are a rich source of vitamin C.

(7) All fruits having 5-petals at the end of a single fruit belong to the rose family. The hip (fruit) and flower are edible and a rich source of vitamin A and C.

(8) The seeds from all grasses are edible.

b. Specific Plants. There are several edible plants found in the western region of the United States and are easy to recognize.

(1) Wild Onions - Complete plant.

(2) Dandelion - Roots and leaves.

(3) Watercress - Complete plant.

(4) Cattail - Root, stalk and stem.

(5) Thistles- Anderson, Bull, and Elk thistle root and flower.

(6) Yampah - Bulb

(7) Juniper Tree - Berries.

(8) Currants - Berries.

5. **MEDICINAL PLANTS (MSVX.02.11c)** A Marine or unit's bid for survival may be complicated by medical problems. Injuries incurred will reduce survival expectancy and the ability to evade.

a. Willow. Willow is a thick forming shrub with clustered stems and very narrow leaves. Its habitat includes wet soils, riverbanks, sandbars, and silt flats.

(1) A fever can be reduced by drinking tea made from the inner bark.

(2) The dried and powdered inner bark can be used to stop severe bleeding.

(3) Diarrhea can be treated by administering a half cup of willow charcoal dissolved in water.

(4) The Willow roots can be mashed and applied to tooth aches.

b. Yarrow. Yarrow is a plant (tall, usually not branched, with many white, slat topped group flowers). Do not confuse Yarrow with Poison Hemlock.

(1) Insect repellent can be made by rubbing a handful of crushed Yarrow flowers and leaves on any exposed skin.

(2) Bleeding can be stopped by placing a Yarrow leaf poultice on the wound.

(3) Relief from minor burns and many rashes, including poison oak and ivy, can be achieved by applying a Yarrow leaf compress to the effected area.

(4) The Yarrow root can be chewed to relieve the pain of a tooth ache or break a fever.

(5) A potent anesthetic can be made by scrubbing fresh Yarrow roots in water to clean them. Once the roots are clean, crush them into a spongy mass and apply gently to the wounded area.

c. Medicinal terms and definitions. The following terms, and their definitions, are associated with medicinal plant use:

(1) Infusion or Tea. The preparation of medicinal herbs for internal or external application. You place a small quantity of a herb in a container, pour hot water over it, and let it steep (covered or uncovered) for 20 minutes before use.

(2) Poultice. The name given to crushed leaves or other plant parts that are applied to a wound or sore either directly or wrapped in cloth. Place plant in gauze or other similar material and fold it so the gauze will hold the plant in place. Put gauze in a larger cloth, about 6"x8", and roll the sides inward. Fold the cloth over without losing the plant. While boiling water, dip the bottom portion of the cloth containing the plant into the hot water by holding the edges. Keep the plant submerged in the boiling water until it becomes saturated. Bring the cloth straight up and with a twisting motion, wring the excess water back into the pot. Apply the poultice to the affected area as soon as it has cooled enough to place on the wound. To be effective, the poultice should be as hot as you can tolerate.

(3) Compress. A compress is made just like a poultice, except it is cold when applied to the wound.

NOTE: Poultice or Compress should be applied for 1 to 24 hours, as needed. When applying a poultice, you may experience a throbbing pain as it draws out the infection and neutralizes toxins. When the pain subsides, the poultice has accomplished its task and should be removed. Apply a fresh poultice as needed until the desired level of healing has been reached.

6. **POISONOUS PLANTS. (MSVX.02.11d)** Rarely will a survivor have an ideal means of killing large game, though there are certain plants that can aid the survivor. The two plants that we will talk about are the Water Hemlock and Monkshood.

   a. Preparing Poisonous Plants. Once a poisonous plant has been located, dig up the root of the plant. The roots of the Water Hemlock and Monkshood generally grow 8 to 10 inches deep. **Extreme** caution must be used when handling the root. **Do not** handle the root without a barrier between your hands and the root. A barrier can be gloves, socks, T-shirt, or even moss. Split the root lengthwise to expose the inside of the root where the toxin is located. With the root split, rub the tip of your spear/arrow inside of the root opening. In a slow controlled manner, work from the bottom of the tip to the top. Once the tip is thoroughly coated, allow the toxin to dry and apply another coat to the tip. Continue to apply coat after coat of toxin, until the root is completely drained of it's toxin. You are now finished and the tip is ready for use.

7. **INSECTS**. Insects are the most abundant life form on earth and are an excellent survival food. They are easy to catch and provide 65-80% protein, compared to 20% for beef. They aren't too appetizing, but personal bias has no place in a survival situation. The focus must remain on maintaining your health.

   a. Insects to avoid. **(MSVX.02.12a)**

      (1) All adults that sting or bite.

      (2) Hairy or brightly colored insects.

      (3) All Caterpillars.

      (4) Insects that have a pungent odor.

      (5) All spiders.

      (6) Disease carriers like ticks, flies, or mosquitoes.

   b. Edible Insects.

      (1) Insect Larvae.

      (2) Grasshoppers.

      (3) Beetles.

      (4) Grubs.

      (5) Ants.

(6) Termites.

(7) Worms.

c. Foraging for Insects. One must be careful not to expend more energy harvesting food than can be replaced. For example, catching insects such as grasshoppers can become frustrating and tiring.

(1) At night grasshoppers climb tall plants and cling to the stalks near the top. They can be picked from the plants in the early morning while they are chilled and dormant.

(2) Dig for worms in damp humus soil, under rocks/logs or look for them on the ground after it has rained.

(3) Carpenter ants are found in dead trees and stumps which can be gathered by hand.

(4) Most other insects can be found in rotten logs, under rocks, and in open grassy areas.

d. Preparing Edible Insects. **(MSVX.02.12b)**

(1) Insects with a hard outer shell have parasites. Remove the wings and barbed legs before cooking.

(2) Drop worms into potable water for at least a half hour. They will naturally purge themselves. You can either cook or eat them raw.

(3) Most other insects can be eaten raw. Cooking insects will improve their taste. If the thought of eating insects is unbearable, grind them into a paste and mix with other foods.

REFERENCE:

1. APM 64-5, Aircrew Survival 1985

2. FM 21-76 Survival 1992

3. Edible Plants, Video

4. Poisonous Plants of California

5. North American Trees

6. Healing Plants & Poisonious, WSI, Chris Janowski, Video

UNITED STATES MARINE CORPS
Mountain Warfare Training Center
Bridgeport, California 93517-5001

MSVX.02.13
2/6/05

STUDENT HANDOUT

SURVIVAL FISHING

TERMINAL LEARNING OBJECTIVE In a survival situation, and given a survival kit, procure fish, in accordance with the references. **(MSVX.02.13)**

ENABLING LEARNING OBJECTIVES

(1) Without the aid of references, list in writing the fishing locations, in accordance with the references. **(MSVX.02.13a)**

(2) Without the aid of references, construct a field expedient hook, in accordance with the references. **(MSVX.02.13b)**

(3) Without the aid of references, conduct fishing, in accordance with the references. **(MSVX.02.13c)**

(4) Without the aid of references, consume a prepared fish, in accordance with the references. **(MSVX.02.13d)**

OUTLINE

1. **FISH**. In a mountainous region, fish are normally an abundant resource. Not only are they a food source, the "left overs" provide an excellent bait for traps and snares.

   a. Fishing locations. **(MSVX.02.13a)** Fishing in mountain streams is generally best done with a hand line. When fishing these streams, always look for these places to fish.

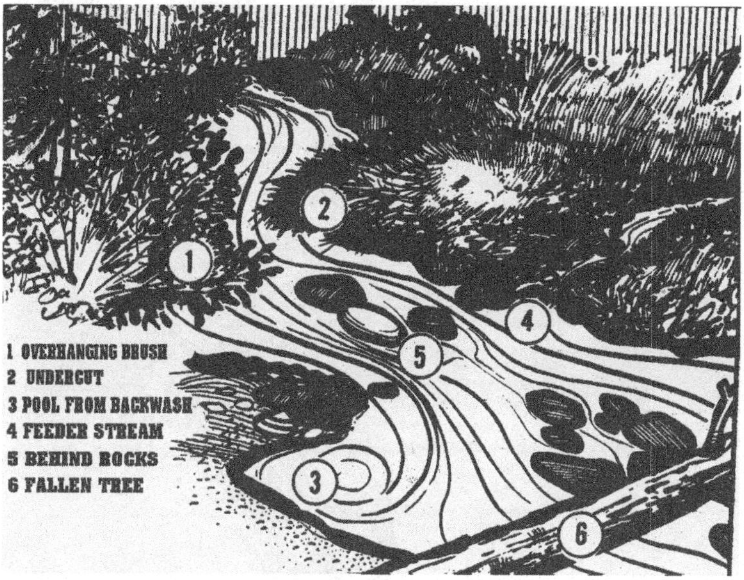

FISHING LOCATIONS

2. **FISHING EQUIPMENT.** Depending upon your location, resources available to you, type of water source, and type and/or size of fish, certain fishing equipment may be needed.

   a. Expedient hooks. **(MSVX.02.13b)** Although hooks should be carried in a survival kit, the survivor should be able to construct additional hooks if the situation arises. Expedient hooks are made to become lodged in the throat of the fish. Below are a few examples.

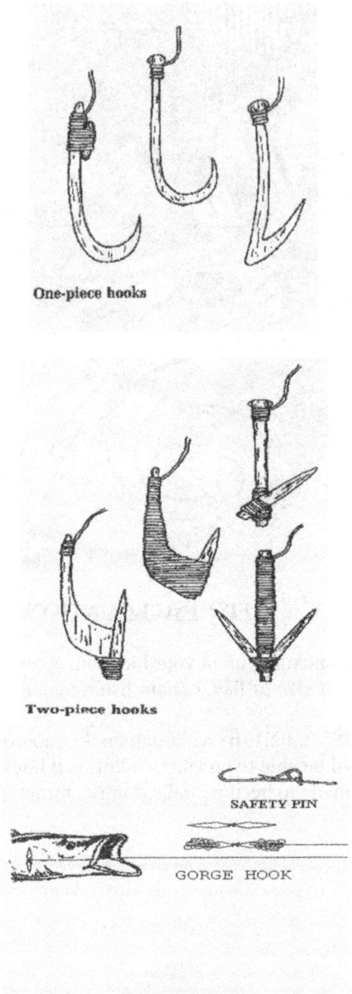

<u>EXPEDIENT HOOKS</u>

b. <u>Fishing Spear</u>. If you are near shallow water (about waist deep) where fish are large and plentiful, you can spear them.

   (1) Cut an 18-24" long straight hardwood sapling, fire harden if green.

   (2) Sharpen one end of the sapling.

(3) Shave two green saplings to serve as prongs.

(4) Carve barbs on the prongs.

(5) Notch main staff to support prongs.

(6) Lash the prongs to the main staff.

(7) Prongs that do not easily flex under the weight of a finger need to be shaved and thinned, prior to lashing.

(8) To spear fish, find an area where fish either gather or where there is a fish run. Place the spear point into the water and slowly move it towards the fish. Then, with a sudden push, impale the fish on the stream bottom. Do not try to lift the fish with the spear. Hold the spear with one hand and grab and hold the fish with the other. Do not throw the spear as you will probably lose it.

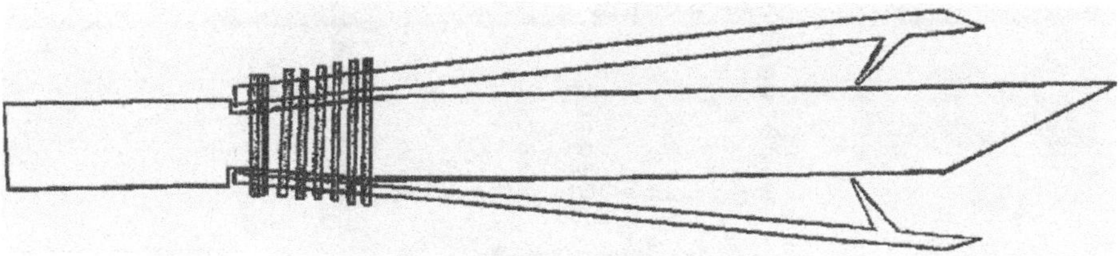

FISHING SPEAR

c. Fish Gaff. A fish gaff is an effective method to procure fish from a concentrated area near the bank or to lift fish out of the water when hooked on a hand line. It is made from a single piece of wood and sharpened on the short end.

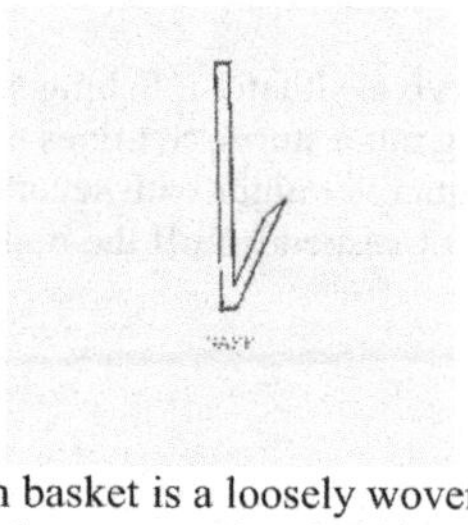

d. Chum Basket. The chum basket is a loosely woven basket that is filled with fish intestines and hung over the waters edge. Within a couple of days, maggots will form and drop into the water, causing the fish to concentrate in the area. The maggots can be used for bait on a hand line.

3. **Fish Traps**. A fish trap can be effective if you have a shallow stream and time to construct it. A basic fish trap is nothing more than a barricade of rocks or sticks across a stream with another barricade using a funnel-type entrance which fish can be driven into but have a difficult time finding their way out. Once fish are trapped between these two barricades they may be speared, clubbed, or grabbed. This can be very effective when certain types of fish are moving in large groups to spawn. This type of trap is very effective in catching fish. A door can be constructed at the mouth of the trap so that excess fish can be kept live until needed.

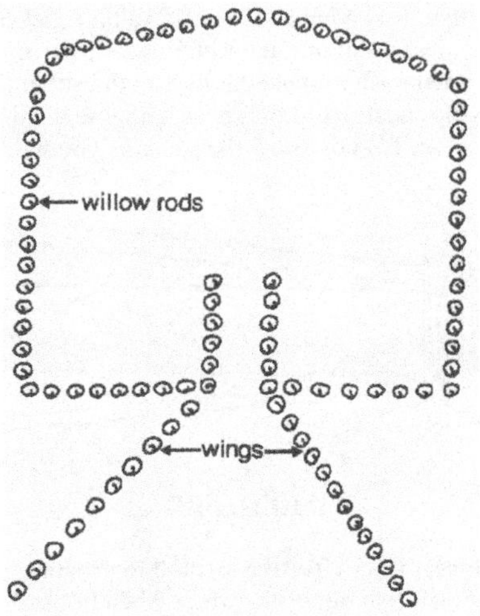

FISH TRAP

4. **FISHING SET LINES**. In a survival situation, fishing sets are means to catch fish while working on other tasks or weathering out a storm. Set lines are an effective method of fishing while conserving energy. Put them out over night with several baited hooks attached. Place them with the hooks either on the bottom or suspended off the bottom, until you have determined where the fish are feeding._

a. Standard Set Lines.

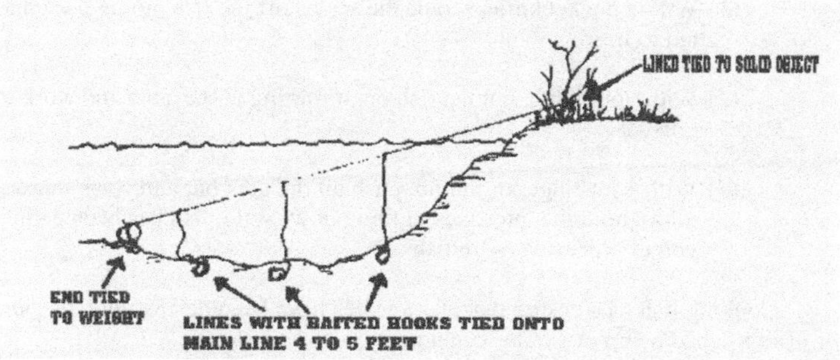

b. Netting. A gill-net is most effective in still water, e.g., a lake (near the inlet and outlet are good locations) or back water in a large stream (for survival don't hesitate to block the stream). Nets can be constructed using the inner cords of parachute shroud lines. Place floats on top and weights on the bottom to keep the net vertical in the water. When ice is on a lake, the fish are inclined to stay deeper. The smaller the mesh, the smaller the fish you can catch, but a small nest will still entangle a large fish. A mesh of 3 inches is a good standard. A gill-net is very time consuming and requires allot of material to construct.

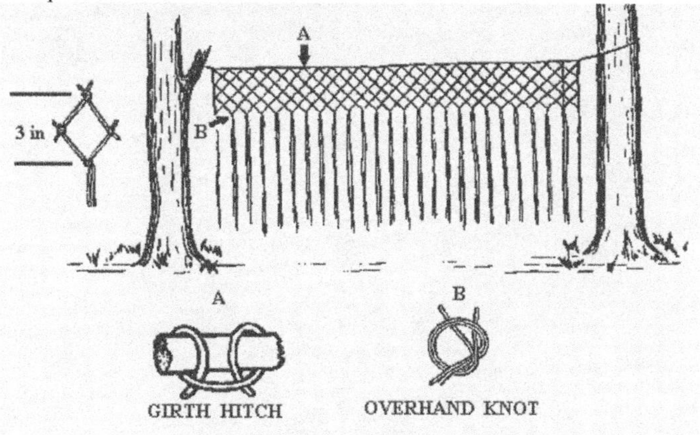

GILL NET

5. **PREPARING FISH FOR CONSUMPTION. (MSVX.02.13d)** Fish may contain many parasites, which if prepared improperly can infect the human body.

a. Cleaning fish.

(1) With a pocket knife, scrape the scales off the fish, going back and forth from tail to head.

(2) With your knife, cut the fish open starting at the anus and work towards the gills.

(3) With your finger or thumb, push all the guts out and wash thoroughly. Look throughout the intestines to find out what the fish has been eating. It may aid you in procuring more fish.

b. Cooking fish. To ensure that all parasites have been destroyed, fish should be boiled in a canteen cup or similar container.

REFERENCE:

1. Larry Dean Olson, Outdoor Survival Guide, 5th Editon, 1990.
2. Chris Janowski, A manual that could save your life, 1989.
3. AF 64-4, Search and Rescue Survival Training, 1985.

UNITED STATES MARINE CORPS
Mountain Warfare Training Center
Bridgeport, California 93517-5001

MSVX.02.14
2/6/05

STUDENT OUTLINE

TRACKING

TERMINAL LEARNING OBJECTIVE In a survival situation, and given a survival kit, and a designated area, identify tracks, in accordance with the references. **(MSVX.02.14)**

ENABLING LEARNING OBJECTIVES

(1) Without the aid of references, describe in writing the types of gaits, in accordance with the references. **(MSVX.02.14a)**

(2) Without the aid of references, describe in writing the tracks of the major animal families, in accordance with the references. **(MSVX.02.14b)**

(3) Without the aid of references and given an animal hide, identify the animal, in accordance with the references. **(MSVX.02.14c)**

(4) Without the aid of references, list in writing the factors that determine track age, in accordance with the references. **(MSVX.02.14d)**

OUTLINE:

1. Basic Terminology. Prior to discussing tracking, some basic terms must be understood by all.

   a. Trails and Runs. In any area, there will be many thoroughfares or trails and runs. Some may be seasonal, while others may be used by many different species. Runs are infrequently or intermittently used thoroughfares that connect trails to specific feeding, bedding, or watering areas. If trails are like highways connecting cities and towns, runs are like streets providing access to the gas stations, supermarkets, and neighborhoods.

   b. Beds and Lays. Beds are frequently used sleeping areas commonly referred to as dens or burrows. These can be found in hollow logs, trees, rock piles, brush piles, grass, thickets, or even out in the open. A lay is an infrequently used resting or sleeping spot. It is rarely used more than once.

c. Rubs. Some rubs are accidental and some are deliberate. Accidental rubs can be in a burrow, on a trail, or over/under a fallen tree across a trail. Deliberate rubs can be when an animal scratches a hard-to-reach spot, or when a deer scrapes its antlers against a tree to remove its velvet.

d. Scratches. They also can be accidental or deliberate. Accidental scratches are left by animals climbing trees or on a log where it left a belly rub. Deliberate scratches can be found at the base of trees where they have reached up and raked their claws downward for any number of reasons. Scratches can also be found in the ground where cats have buried scat, squirrels have cached nuts, or animals are digging at a scent.

e. Transference. Transference is the removal of material from one area onto another. Transference can occur when walking along a muddy stream bank and crossing a log. The mud left on the log is considered transference.

f. Compression. Compression is the actual flattening of the soil or snow pack. It is caused by the pressing down or leveling of soil, sand, stones, twigs, or leaves by the weight of the body. Compression is more likely to be found in frozen, hard, dry, sandy conditions where there is no moisture to hold a clear and lasting imprint.

g. Disturbance. Disturbance is the eye-catching effect of unnatural patterns.

h. Gait. **(MSVX.02.14a)** A gait is generally the way an animal moves. Gaits are very critical in the identification of animal tracks. Although certain gaits are more indicative of certain animals, they may (depending on the circumstances) modify or alter their gait to another style.

(1) Diagonal Walker. Normal pattern for all predatory animals, which includes all dogs, cats, and hoofed animals.

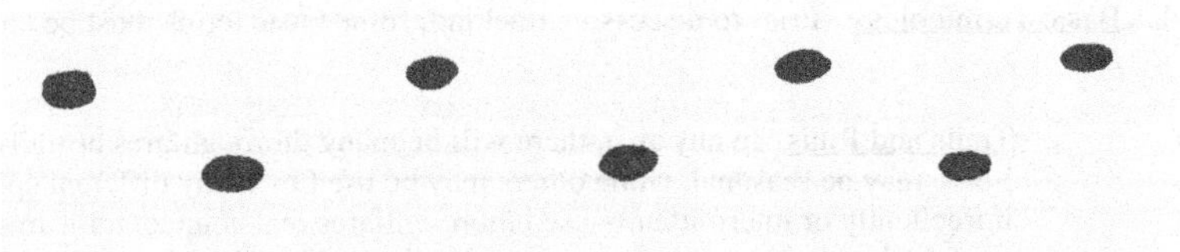

(2) Pacers. Normal pattern for all wide-bodied animals such as bears, raccoon, opossum, skunk, wolverine, badger, beaver, porcupine, muskrat, and marmot. Instead of moving opposite sides of the body at the same time like diagonal walkers, they find it easier to move both limbs on one side of the body at the same time.

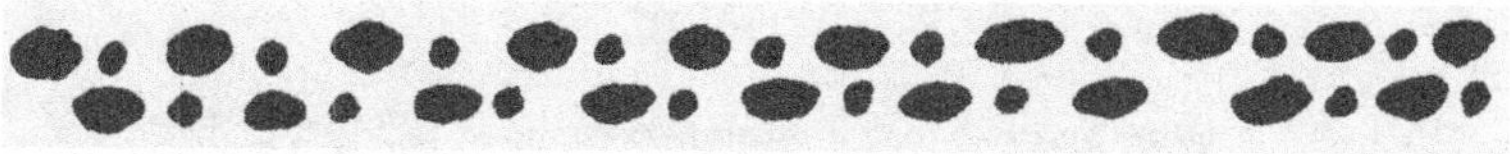

(3) Bounders. Normal pattern for most of the long-bodied, short-legged weasel family such as marten, fisher, and mink. Bounders walk by reaching out with the front feet and bringing the back feet up just behind them.

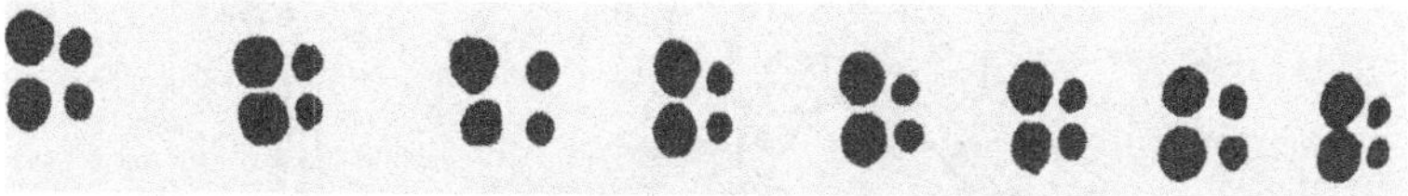

(4) Gallopers. Normal pattern for rabbits, hares, and rodents (except wide-body beaver, muskrat, marmot, and porcupine). As these animals move, they push off with their back feet, hit with their front feet, and bring their back feet into position. Tree dwelling gallopers will land with their front feet side by side, while ground dwelling gallopers will land with the front feet at a diagonal.

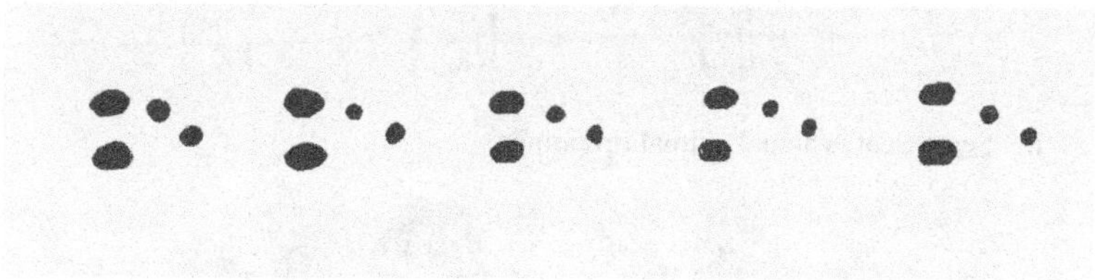

i. Gnawing. All animals will chew on vegetation; some as a food source, while predators need certain vitamins. Gnawings can be on trees (cambium layer) or on vegetation.

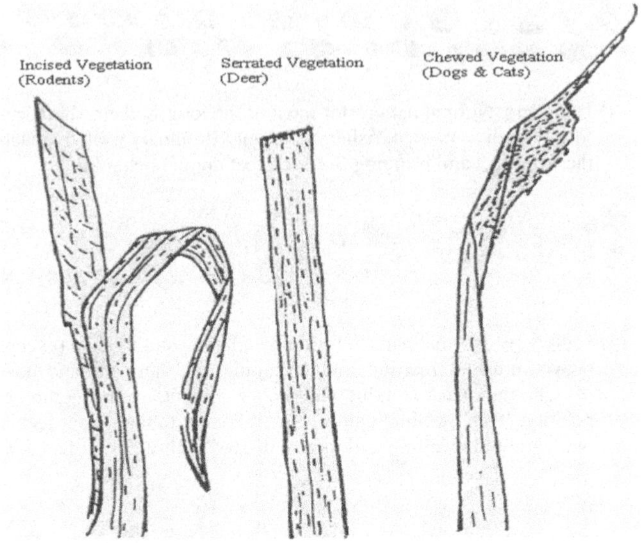

j. Scat. Scat is actual animal droppings.

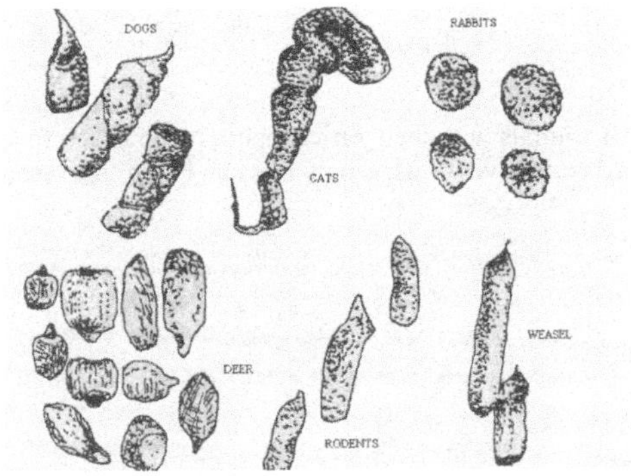

k. Sign. Any disturbance of the natural condition which reveals the presence or passage of animals, persons, or things. Examples of sign include stones that have been knocked out of their original position, overturned leaves showing a darker underside, sand deposited on rocks, drag and scuff marks, displaced twigs, and scuff marks on trees.

l. Spoor. The actual track or trail of an animal which can be identified as to size, shape, type and pattern. This word is generally interchanged with track. Spoor is broken down into two segments; aerial and ground.

2. **<u>Reading Spoor</u>**. Unless a clearly visible ground spoor is readable, interpretations must be made in order to determine "what animal made this?" Prior to ever attempting to read spoor, one must be thoroughly knowledgeable about what animals are in the area. The first step is to look at the gait. This will generally narrow down the species. The next step is determined which animal family the track belongs to. **(MSVX.02.14b)**

   a. <u>Cat Family</u>. Bobcat, Lynx, and Mountain Lion (Cougar). 4 toe pads, no visible claw print. It moves with a sense of purpose and direct registers its paws. Its heel pad is much more defined than one from a dog.

   b. <u>Dog Family</u>. Fox, Coyote, and Wolves. Visible claw print, 4 toe pads. No sense of purpose, except fox, which steps like a cat and likes depressions. Heel pad is much rounder.

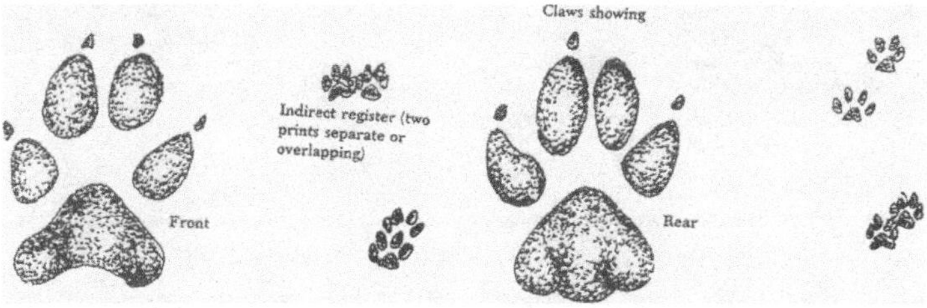

   c. <u>Rabbit Family</u>. The main difference between rabbits and hares (which include the jackrabbit) is that rabbits are born almost hairless and with eyes closed, while hares are born with a thick coat of fur, open eyes, and an ability to run very soon afterwards. They have four toes with relatively enormous hind feet as compared to their front.

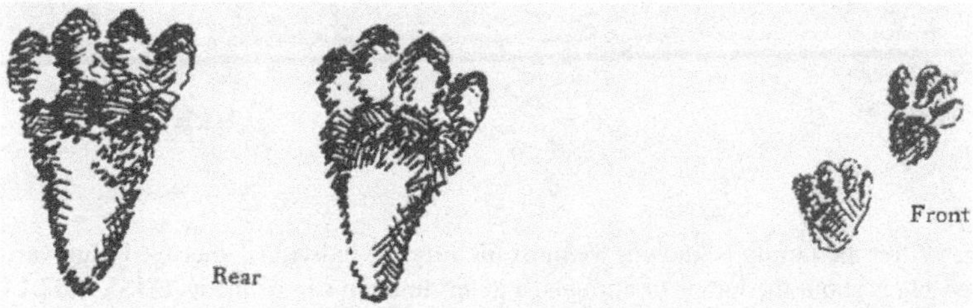

   d. <u>Rodent Family</u>. Voles, Mice, Rats, Chipmunks, Squirrels, Woodchucks, Muskrats, & Beaver. Track size varies greatly because of the different species, but one fact remains, all have 5 toe prints on their rear feet, while having 4 toe prints on their front feet.

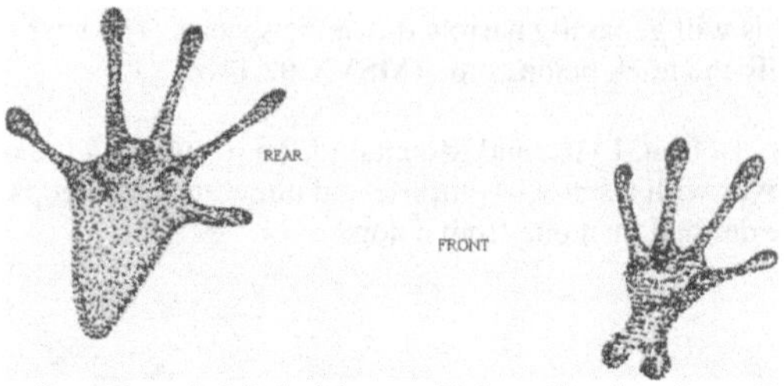

e. Weasel Family. Weasel, Mink, Marten, Fisher, Skunks, Otter, Badger, Porcupine, & Wolverine. All have 5 toe prints.

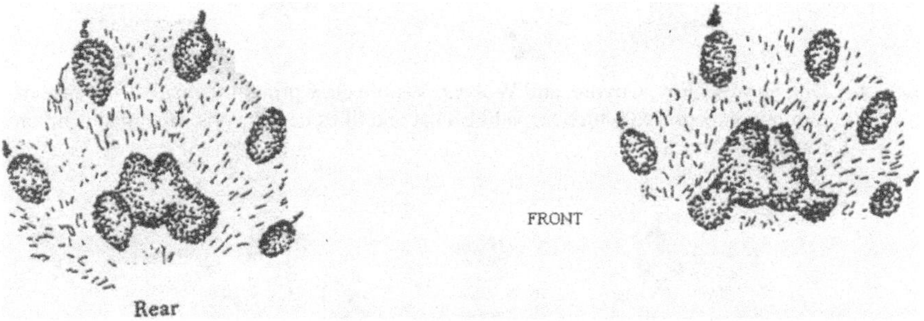

f. Raccoon, Opossums, & Bear. All have 5 toe prints while looking like a baby's hand print.

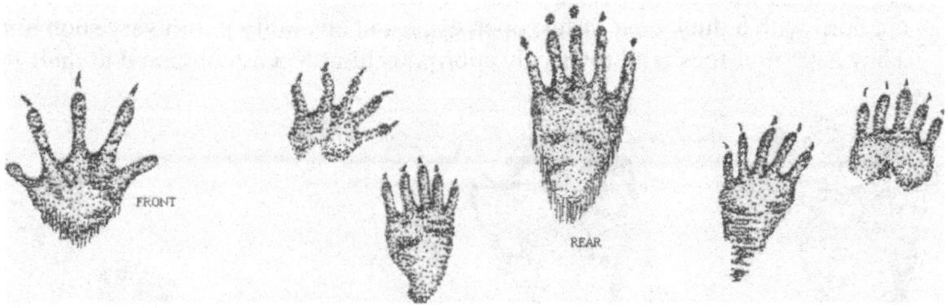

g. After the family is known, we must identify the individual species. Using various clues about the habits of animals, a determination can be made. **(MSVX.02.14c)**

(1) If the tracker is educated on the behavior and habits of animals, he can determine the individual species. This information can be used for better employment of traps and snares. The following is an example.

(a) Walking along a creek bank, you notice a set of tracks that have five toe prints for both front and back feet with visible claw prints. The information tells you the prints belong to the weasel family. The tracks have a bounding gait pattern, which eliminates wide-bodied animals such as badger, skunk, porcupine, and wolverine. Because the tracks are approximately the size of a dime, you have eliminated marten and fisher. The tracks are following the stream bank for some distance, stopping at small holes along the bank's edge. Knowing that weasels like grassy meadows, you can determine that the track is probably made by a mink.

3. **Age Determination**. **(MSVX.02.14d)** It is very critical to be able to determine track age. Each area and climate will vary in the effects of aging tracks, so practice, experimentation, and experience is vital in that area. The following factors deteriorate all tracks and must be factored.

a. Weather - Last snow or rain, fog, and dew.

b. Sun.

c. Wind.

d. Soil content - hard, sandy, firm, or moist.

e. Track Erosion. All tracks will erode over a given period of time. The following time table can be used as a *general* guideline.

(1) Minutes-Top edges begin to dry.

(2) 24 hours-Top edges begin to erode.

(3) 48 hours-"S" curve is seen.

(4) 72 hours-Pock marks from rain or dew may be seen. Track is almost flat. Debris may fill the track.

f. Aging Scat. All scat dries on the inside first. Therefore, relatively wet scat on the outside could be old. The only way to determine the age is by analyzing the inside. When assessing scat, care must be taken to avoid the possibility of contracting disease.

4. **Tracking**.

a. The best time to track is early in the morning or late in the afternoon due to the height of the sun to cast shadows. When reading spoor, always place it between yourself and the sun.

b. Do not move past the last sign until you have found the next sign, this is called “sign cutting” and will be discussed later. In training, always try to find every track.

c. Once the initial track is found, completely document and sketch it for future reference. This sketch will prevent you from following the wrong track later on. Record the following information.

    (1) Shoe length.

    (2) Sole width.

    (3) Heel width.

    (4) Sole length.

    (5) Heel length.

    (6) Document Quadrants:
        -Wear patterns
        -Tread patters

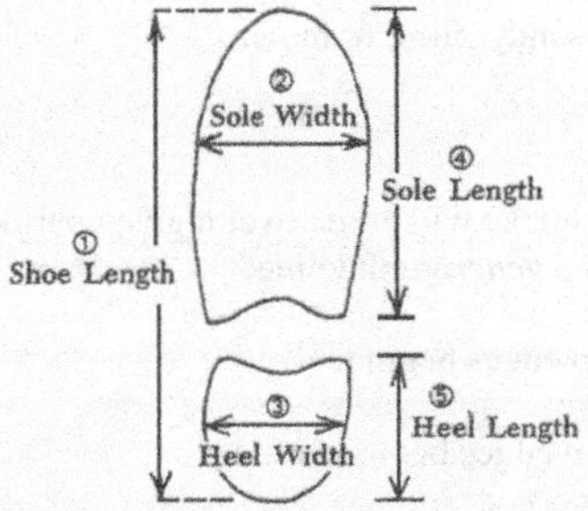

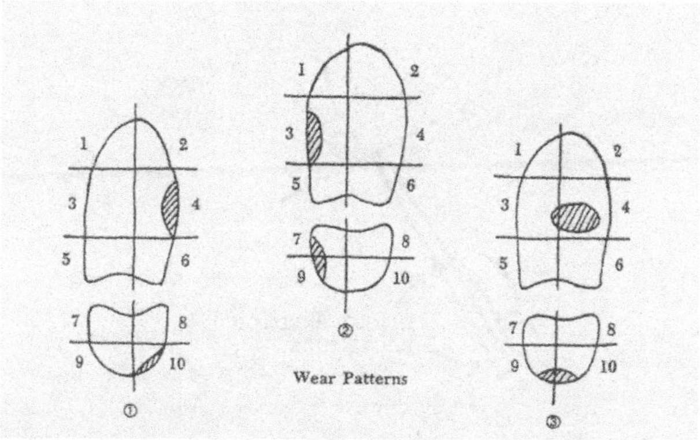

d. Attempting to find each and every track is a slow and tedious task. The best method to assist you is with the use of a **tracking stick**.

(1) Materials: a 4 foot straight pole
2 rubber bands or similar material

(2) Locate 2 consecutive tracks.

(3) Place the tip of the stick on the edge of the heel on the lead track.

(4) Place a rubber band to mark the toe of the rear track.

(5) Place the second rubber band to mark the heel.

(6) Move the tracking stick to the rubber band markings of the lead track and sweep the tip of the stick until the next track is found.

(7) Repeat the process.

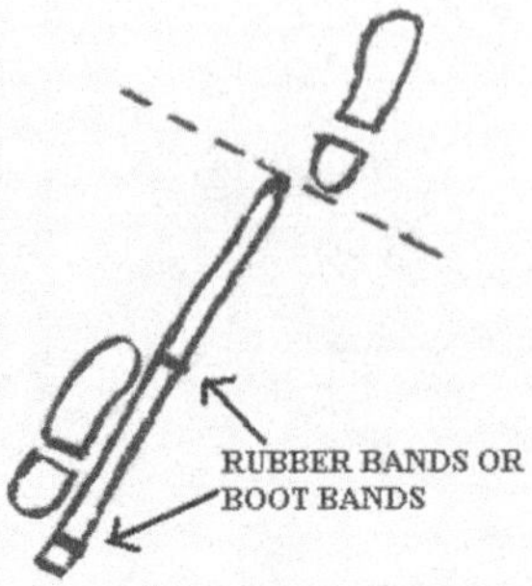

e. Tracking Teams. If tracking teams are available, the tracking process can be sped up by "sign cutting".

(1) All tracking teams (minimum of two) must document and sketch the initial track.

(2) The initial team continues to track until another tracking team has positively found the same track further ahead on the trail. The last track is marked for future reference.

(3) The second team assumes the responsibility of locating each track until they have been radioed by the first team that "they have found the track". This is "sign cutting".

(4) "Sign cutting" is accomplished by making large sweeping arcs ahead of the primary tracking team until the track is located at which time the teams change roles by "leap frogging".

(5) If the track is lost or misidentified, the teams will move back to the last marked track.

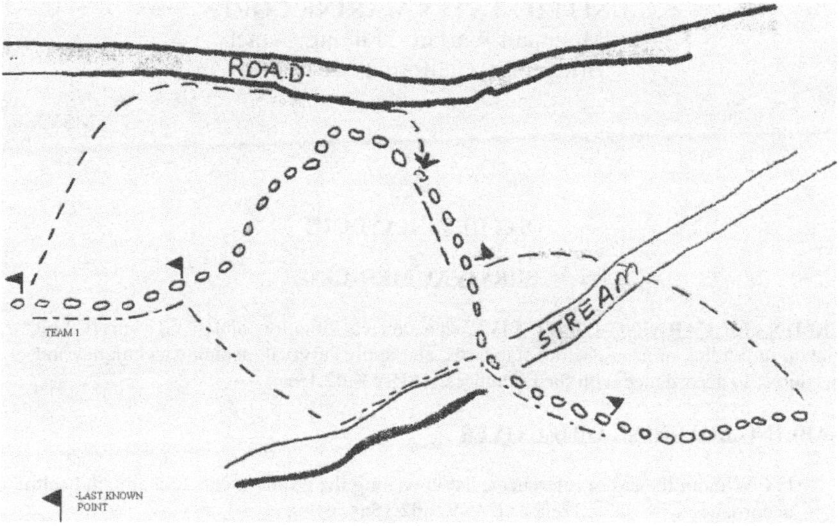

HYPOTHETICAL SEARCH

f. Delaying a tracker or tracking teams. If you are being tracked, your primary concern is to gain as much distance as possible between yourself and the tracker. The more distance you gain, the more time to enable you to create devices to discourage a tracker.

(1) Locate and dig poisonous plants. Trackers know the vegetation in their area. If they have seen poisonous plants dug along the trail, they will use more caution and slow their pace, whether you have used them or not.

(2) Create simple pathguards along your trail. An experienced tracker will not pick up things along the trail because of the possibility of being "booby trapped". If he notices possible traps, he will use more caution and slow his pace.

(3) Use caution when moving along the trail. When traveling, make it difficult for the tracker to find your tracks. Padded over boots are extremely effective to reduce the "signature" left by combat boots.

REFERENCE:

1. David Scott Donelan, Tactical Tracking Operation, 1998.
2. Tom Brown, Field Guide to Nature Observation and Tracking, 1983.

UNITED STATES MARINE CORPS
Mountain Warfare Training Center
Bridgeport, California 93517-5001

MSVX.02.15
2/6/05

STUDENT HANDOUT

SURVIVAL MEDICINE

TERMINAL LEARNING OBJECTIVE In a survival situation, and given a survival kit, maintain individual or group health standards, and apply survival medicine techniques and procedures, in accordance with the references. **(MSVX.02.15)**

ENABLING LEARNING OBJECTIVES

(1) Without the aid of references, list in writing the requirements to maintain health, in accordance with the reference. **(MSVX.02.15a)**

(2) Without the aid of references, list the in writing the environmental injuries, in accordance with the reference. **(MSVX.02.15b)**

(3) Without the aid of references, define in writing the definition of hypothermia, in accordance with the reference. **(MSVX.02.15c)**

(4) Without the aid of references, list in writing the general considerations for medevac procedures, in accordance with the reference. **(MSVX.02.15d)**

(5) Without the aid of references, execute preventive measures against wildlife diseases, in accordance with the reference. **(MSVX.02.15e)**

OUTLINE

1. REQUIREMENTS TO MAINTAIN HEALTH. Maintenance of health is the first step in preventing injuries. The three requirements are: **(WSVX.02.15a)**

A. WATER

(1) A person can not survive without water for more than a few days. Your body loses water through normal body processes (sweating, urinating, and breathing). During normal activity the kidneys excrete 1 to 2 quarts of water per day and a person evaporates . 5 to quart per day. Other factors, such as heat exposure, cold exposure, intense activity, high altitude, burns, or illness, can cause your body to lose more water. Water intake is critical in preventing illness.

B. FOOD

(1) The body can live several weeks without food. However, without an adequate supply to stay healthy your mental and physical capabilities will deteriorate rapidly. Food supplies the body with the necessary nutrients and energy to survive.

(2) Food sources are plants, animals, and fish.

(a) Fiber. Fiber prevents irritable bowel syndrome. In the Falklands campaign the British had a major constipation and diarrhea problem. This was largely caused by dehydration and a low fiber diet. Grasses and pine needles are a good source of dietary fiber.

(b) Vitamins. Vitamins are essential to metabolic functioning of the body and cold weather compounds this function. Our bodies cannot make vitamins so we must provide them in our diet. Most edible plant life contains many different vitamins. Associated illnesses from long term deficiency are Scurvy (vitamin C) a physical disease and Beriberi (vitamin B1) a mental disease. Vitamins can be found in the cambium layer of trees, pine needles, and stinging nettle.

(c) Minerals. The mineral that we are primarily concerned with is Iron. Iron deficiency causes a 9% decrease in heat energy production. Iron acts as a thermo regulator. Consuming only 1/3 RDA of iron results in a 29% greater heat loss during cold exposure. Animal blood, dandelions, stinging nettle, and marrow provide the major source of iron. Ensure these foods are properly prepared.

(d) Calories. To produce energy, the body uses calories. Proteins, fats, or carbohydrates produce calories. Of these three, certain ones produce better energy than others do. Animal meat is an excellent source for caloric intake, although nuts from pinecones can supplement it.

1. Protein. Proteins are a reparative food of complicated molecules composed of chains of amino acids. There are numerous kinds of amino acids which cannot be synthesized by the body and thus must be consumed in the diet. A pure protein diet can cause fatalities in 3-8 weeks from *Rabbit Starvation*, a term used for living on a relatively fat free rabbit diet. Protein can be found in dandelions, nuts, and meats.

2. Fat. Fats serve as the main storage form of energy. Fats produce energy and heat. Fats are best obtained from bone marrow, liver, or the stomach portion of fish

3. Carbohydrates. Carbohydrates are known as the quick energy food. They produce lots of heat. They are stored in the liver and muscles. These organs are not large and can be markedly depleted by fasting for as short as 24 hours. Cattails, nuts are a source for carbohydrates.

C. PERSONAL HYGIENE

(1) Cleanliness is an important factor in preventing infection and disease. It becomes even more important in a survival situation.

(2) The areas to pay special attention to are the feet, hands, armpits, crotch, and hair. Visual and physical inspections should be conducted daily. Hand and finger nails should be kept as clean as possible to prevent infection of mucous membrane.

(a) Soap (lye) can be made from animal fat and ashes.
(b) Sun bath

(3) Teeth are another important area to keep clean. Brush your teeth each day either with a toothbrush, or if you don't have one, make a chewing stick.

(a) A chewing stick is made out of a twig about 6 to 8 inches long. Chew one end of the stick to separate the fibers. Now you can brush your teeth.

2. FIVE WAYS THE BODY LOSES HEAT

A. Radiation: is direct heat loss from the body to its surroundings. If the surroundings are colder than the body, the net result is heat loss. A nude man loses about 60% of his total body heat by radiation. Specifically, heat is lost in the form of infrared radiation. Infrared targeting devices work by detecting radiant heat loss.

B. Conduction: is the direct transfer of heat from one object in contact with a colder object.

(1) Most commonly conduction occurs when an individual sits or rests directly upon a cold object, such as snow, the ground, or a rock. Without an insulating layer between the Marine and the object (such as an isopor mat), one quickly begins to lose heat. This is why it's important to not sit or sleep directly on cold ground or snow without a mat or a pack acting as insulation.

C. Convection: is heat loss to the atmosphere or a liquid.

Air and water can both be thought of as "liquids" running over the surface of the body. Water or air, which is in contact with the body, attempts to absorb heat from the body until the body and air or water is both the same temperatures. However, if the air or water is continuously moving over the body, the temperatures can never equalize and the body keeps losing heat.

D. Evaporation . Heat loss from evaporation occurs when water (sweat) on the surface of the skin is turned into water vapor. This process requires energy in the form of heat and this heat comes from the body.

(1) This is the major method the body uses to cool itself down. This is why you sweat when you work hard or PT. One quart of sweat, which you can easily produce in an hour of hard PT, will take about 600 calories of heat away from the body when it evaporates.

E. Respiration. When you inhale, the air you breathe in is warmed by the body and saturated with water vapor. Then when you exhale, that heat is lost. That is why breath can be seen in cold air. Respiration is really a combination of convection (heat being transferred to moving air by the lungs) and evaporation, with both processes occurring inside the body.

3. **PHYSICAL RESPONSES TO HEAT.** When the body begins to create excess heat, it responds in several ways to rid itself of that heat.

A. Initially, the blood vessels in the skin expand, or dilate. This dilation allows more blood to the surface where the heat can more easily be transferred to the surroundings.

B. Soon afterwards, sweating begins. This contributes to heat loss through convection and evaporation.

4. **PHYSICAL RESPONSES TO COLD.** Almost the opposite occurs as with heat.

A. First, blood vessels at the skin surface close down, or constrict. This does two things:

(1) Less blood goes near the surface of the body so that less heat is lost to the outside.
(2) More blood goes to the "core" or the center of the body, to keep the brain, heart,lungs, liver, and kidneys warm. This means fingers and toes tend to get cold.

B. If that is not enough to keep the body warm, the next step is shivering. Shivering is reflexive regular muscular contractions, this muscular activity causes heat production. As mentioned before, shivering can only last for a short time before exhaustion occurs. With shivering you will either warm up, as usually occurs, or continue to get colder and start to become hypothermic. Hypothermia will be discussed later.

5. <u>ENVIRONMENTAL INJURIES</u> **(MSVX.02.15b)** Are cold weather injuries, dehydration, and altitude related illnesses.

A. <u>**COLD WEATHER INJURIES**</u>:

1) **Hypothermia (MSVX.02.15c)** is the state when the body's core temperature falls to 95 degrees Fahrenheit or less. **It is the number <u>one</u> killer of people in a survival situation**. A common belief that extremely cold temperatures are needed for hypothermia to occur is not true. Most cases occur when the temperature is between 30 and 50 degrees Fahrenheit. This is the normal temperature range at MWTC, except during the harshest of winter.

a) <u>Causes of Hypothermia.</u> The ways in which the body generates and loses heat has been discussed earlier. Quite simply, hypothermia occurs when heat loss from the body exceeds the body's ability to produce heat. Contributing factors include:

- Ambient temperature. Outside air temperature.
- Wind chill. This only affects improperly clothed individuals.
- Wet clothing.
- Cold water immersion.
- Improper clothing.
- Exhaustion.
- Alcohol intoxication, nicotine and drugs such as barbiturates and tranquilizers.
- Injuries. Those causing immobility or major bleeding, major burn and head trauma.

<u>b) Signs and symptoms of Hypothermia</u>

- The number one sign to look for is altered <u>mental status;</u> that is, the brain is literally getting cold. These signs might include confusion, slurred speech, strange behavior, irritability, impaired judgment, hallucinations, or fatigue.

- As hypothermia worsens, victims will lose consciousness and eventually slip into a coma.

- Shivering. Remember that shivering is a major way the body tries to warm itself early on, as it first begins to get cold. Shivering stops for 2 reasons:

- The body has warmed back up to a normal temperature range.

- The body has continued to cool. Below 95°F shivering begins to decrease and by 90°F it ceases completely.

- Obviously, continued cooling is bad. So if a Marine with whom you are working, who was shivering, stops shivering, you must determine if that is because he has warmed up or continued to cool.

- A victim with severe hypothermia may actually appear to be quite dead, without breathing or a pulse. However, people who have been found this way have been successfully "brought back to life" with no permanent damage. So remember, ***you are not dead until you are warm and dead***.

c) Prevention of Hypothermia

- Obviously, prevention is always better (and much easier) than treatment.

- Cold weather clothing must be properly warm and cared for.

- Keep your clothing as dry as possible.

- If your feet are cold, wear a hat. Up to 80% of the body's heat can escape from the head.

- Avoid dehydration. Drink 6 - 8 quarts per day.

- Eat adequately.

- Avoid fatigue and exhaustion.

- Increase levels of activity as the temperature drops. Do not remain stationary when the temperature is very low. If the tactical situation does not permit moving about, perform isometric exercises of successive muscles.

- Use the buddy system to check each other for signs/symptoms of hypothermia.

d) Treatment of Hypothermia.

- Make the diagnosis.
- Prevent further heat loss.
- Remove the victim from the environment (i.e., into a shelter or snow cave).
- Insulate the victim.
- Rewarm the victim by:
  *Zip two sleeping bags together.
  *Pre-warm the bag by a stripped Marine.
  *Place the victim in the bag with 2 stripped Marines inside on both sides of the victim.
- Medevac if possible.

e) Other Points to Remember.

- Fluids. If the victim is mildly hypothermic, he may be given hot wets. Otherwise give him nothing by mouth.

- Avoid, if possible, excessive movement of the victim, as his heart may stop beating if it is jarred.

- Major Wounds. Apply first aid to major wounds first, before attempting to re-warm the victim. Re-warming a victim who has bled to death does little good.

- Never give alcohol to hypothermia victims.

- Even after you have started re-warming a victim, he must be constantly monitored. Don't forget about him.

2) **FROSTBITE** Frostbite is the actual freezing of tissues. When in a survival situation, rewarming a severe frostbitten area will not help. It is best to wait for rescue and medical attention.

a) Prevention of Frostbite. Frostbite is an entirely preventable injury.

- Dress in layers. Keep comfortably cool. If you begin to become uncomfortable, add layers.
- Keep clothes dry. If clothing (especially socks and gloves) become wet, change them. This may mean you have to change sock 4-5 times a day.

- Dress properly. If the wind is blowing, wear the correct protective layer.
- Avoid dehydration. When dehydrated, the amount of blood available to warm your fingers and toes goes down, increasing the risk of frostbite.
- Avoid Starvation. Remember - Food is Fuel - and the body uses that fuel to make heat.
- **Leadership must ensure that preventive measures are taken.**

b) <u>Signs and Symptoms of Frostbite</u>.

- Ears, nose, fingers and toes are affected first.
- Areas will feel cold and may tingle leading to....
- Numbness which progresses to...
- Waxy appearance with skin stiff and unable to glide freely over a joint.

c) <u>Treatment of Frostbite</u>. Frostbite is classified into three different degrees: Frosting, Superficial Frostbite, and Deep Frostbite.

- Frosting will revert to normal after using the technique of body heat rewarming.
- Hold the affected area, skin to skin for 15 minutes.
    *Rewarm face, nose, and ears with hands.
    *Rewarm hands in armpits, groin or belly.
    *Rewarm feet with mountain buddy's armpits or belly.
- If affected area cannot be rewarmed in 15 minutes, Superficial Frostbite or Deep Frostbite is suspected.

Do not attempt to further rewarm
Splint the affected area.
Protect the affected area from further injury.
Medevac as soon as possible.
DO NOT RUB ANY COLD INJURY WITH SNOW.
Do not massage the affected area.
Do not rewarm with stove or fire: a burn injury may result.
Loosen constricting clothing.
Avoid tobacco products.

d) <u>Treatment of Superficial or Deep frostbite</u>. Any frostbite injury, regardless of severity, is treated the same – evacuate the casualty and re-warming in the rear. Unless the tactical situation prohibits evacuation or you are in a survival situation, *no consideration should be given to re-warming frostbite in the field.* The reason is something-called freeze – thaw – re-freeze injury.

- Freeze – Thaw – Re-freeze injury occurs when a frostbitten extremity is thawed out, then before it can heal (which takes weeks and maybe months) it freezes again. This has devastating effects and greatly worsens the initial injury.
- In an extreme emergency it is better to walk out on a frostbitten foot than to warm it up and then have it freeze again.

- Treat frozen extremities as fractures - carefully pad and splint.

- Treat frozen feet as litter cases.

- Prevent further freezing injury.

- Do not forget about hypothermia. Keep the victim warm and dry.

- Once in the rear, a frostbitten extremity is re-warmed in a water bath, with the temperature strictly maintained at 101°F - 108°F.

3) SNOW BLINDNESS

a. Definition. Sunburn of the cornea.

b. Causes of Snow Blindness. There are two reasons Marines in a winter mountainous environment are at increased risk for snow blindness.

- High altitude. Less ultraviolet (UTV) rays are filtered out, UV rays are what cause snow blindness (as well as sunburn). So at altitude, more UV rays are available to cause damage.

- Snow. The white color of snow reflects much more LTV rays off of the ground and back into your face.

c. Signs and Symptoms of Snow Blindness.

- Painful eyes.

- Hot, sticky, or gritty sensation in the eyes, like sand in the eyes.

- Blurred vision.

- Headache may be severe.

- Excessive tearing.

- Eye muscle spasm.

- Bloodshot eyes.

d. Prevention of Snow Blindness. Prevention is very simple. Always wear sunglasses, with UV protection. If sunglasses are not available, then field expedient sunglasses can be made from a strip of cardboard with horizontal slits, and charcoal can be applied under the eyes to cut down on reflection of the sun off the snow.

e. Treatment of Snow Blindness.

- Evacuation, when possible.
- Patch the eyes to prevent any more light reaching them.
- Wet compresses, if it is not too cold, may help relieve some of the discomfort.
- Healing normally takes two days for mild cases or up to a week for more severe cases.

4) **TRENCHFOOT / IMMERSION FOOT**

a) Definition. This is a cold - wet injury to the feet or hands from prolonged (generally 7 - 10 hours) exposure to water at temperatures above freezing.

b) Causes of Trench foot/Immersion Foot. The major risk factors are wet, cold and immobility.

c) Signs and Symptoms of Trench foot/Immersion Foot.

- The major symptom will be pain. Trench foot is an extremely painful injury.
- Trench foot and frostbite are often very difficult to tell apart just from looking at it. Often they may both be present at the same time. Signs include:
- Red and purple mottled skin.
- Patches of white skin.
- Very wrinkled skin.

- Severe cases may leave gangrene and blisters.

- Swelling.

- Lowered or even absent pulse.
- Trench foot is classified from mild to severe.

d) Prevention of Trench foot/Immersion Foot is aimed simply at preventing cold, wet and immobile feet (or hands).

- Keep feet warm and dry.

- Change socks at least once a day. Let your feet dry briefly during the change, and wipe out the inside of the boot. Sock changes may be required more often.

- Exercise. Constant exercising of the feet whenever the body is otherwise immobile will help the blood flow.

e) Treatment of Trench foot/Immersion Foot.

- All cases of trench foot must be evacuated. It cannot be treated effectively in the field.

- While awaiting evacuation:
  The feet should be dried, warmed, and elevated.
  The pain is often severe, even though the injury may appear mild; it may require medication such as morphine.

- In the rear, the healing of trench foot usually takes at least two months, and may take almost a year. Severe cases may require amputation. *Trench foot is not to be taken lightly.*

B. DEHYDRATION

1. **Dehydration** is a deficit of total body water. Dehydration will compound the problems faced in a survival situation. Dehydration is the **second leading cause of all deaths in a survival situation**.

   a) Symptoms. When dehydrated, the following signs and symptoms will appear:

   - Headache and nausea.

   - Dizziness and fainting.

   - Cramps, both abdominal and extremity.

- Weakness and lethargy.
- Dark urine with a very strong odor.

b) Prevention. Prevention is the key to prevent dehydration. The following are basic guidelines for the prevention of dehydration:

- Always drink water when eating. Water is used and consumed as a part of the digestion process. If you have plenty of food but no water – Do not eat until a source of water can be found.
- Conserve energy. Pace yourself.
- Drink 6-8 quarts of water per day when available. In other words, continually drink through out the day. Don't wait until you are dehydrated.
- Monitor the color of your urine.
- Don't rely on thirst as an indicator.

2. **Heat related illnesses**. The following illnesses will appear from dehydration:

    1. Heat syncope. Heat syncope is feinting due to vaso-dilation from the heat.
    2. Heat exhaustion. Heat exhaustion occurs when body salt losses and dehydration from sweating are so severe that a person can no longer maintain adequate blood pressure. Heat exhaustion can lead to heat stroke.

        a) Symptoms include; headaches, nausea, dizziness, fatigue, and fainting.

    3. Heat stroke. Heat stroke is a failure of the body's cooling mechanisms that rid the body of excessive heat build up.

        a) Signs and symptoms
        Symptoms are the same as heat exhaustion . The signs include delirious or coma, pinpoint pupils, flushed skin, sweating may or may not be present.

        b) Heat cramps. Heat cramps are painful spasms of skeletal muscle as a result of body salt.

c) All of these illnesses can be detrimental to your survival. Dress properly, rest and **adequate water intake** can help prevent these illnesses.

C. ALTITUDE RELATED ILLNESSES:

1. Acute Mountain Sickness. Acute Mountain Sickness (AMS) is a self-limiting illness due to the rapid exposure of an unacclimatized individual to high altitude (i.e., helicopter crash on a mountain). Approximately 25% of individuals who ascend rapidly to 8,000 – 9,000 feet will develop AMS. Virtually, all un-acclimatized persons who rapidly ascend to 11,00 – 12,000 feet will develop AMS.

   (a) Signs and symptoms include; apathy, dizziness, easily fatigued, nausea, decreased appetite, headache. Can be misdiagnosed as dehydration. If adequate fluid intake is maintained and headache still persists rule out dehydration.

2. HACE. HACE or High Altitude Cerebral Edema is swelling of the brain

   (a) Sign and symptoms are similar to AMS and accompanied by bizarre behavior, hallucinations, confusion, and severe cases – coma.

3. HAPE. High Altitude Pulmonary Edema is the filling of the lungs with fluid.

(a) Signs and symptoms include; persistent cough with pink frothy sputum, shortness of breath, disorientation, fainting, cool and clammy skin, blue lips

   (1) Treatment **Descend, Descend, and Descend.** HACE and HAPE can result in death.

   (2) Prevention Gain elevation slowly. 10,000 feet move 1000 feet per day over 14,000 move no faster than 500 – 1,000 feet per day. Rest and acclimatize your body.

A. CARBON MONOXIDE POISONING

1. Definition. Carbon Monoxide (CO) is a heavy, odorless, colorless, tasteless gas resulting from incomplete combustion of fossil fuels. CO kills through asphyxia even in the presence of adequate oxygen, because oxygen-transporting hemoglobin has a 210 times greater affinity for CO than for oxygen. What this means is that CO replaces and takes the place of the oxygen in the body causing Carbon Monoxide poisoning.

2. Signs/Symptoms. The signs and symptoms depend on the amount of CO the victim has inhaled. In mild cases, the victim may have only dizziness, headache, and confusion; severe cases can cause a deep coma. Sudden respiratory arrest may occur. The classic sign of CO poisoning is cherry-red lip color, but this is usually a very late and severe sign, actually the skin is normally found to be pale or blue.

CO poisoning should be suspected whenever a person in a poorly ventilated area suddenly collapses. Recognizing this condition may be difficult when all members of the party are affected.

3. Treatment. The first step is to immediately remove the victim from the contaminated area.

a) Victims with mild CO poisoning who have not lost consciousness need fresh air and light duty for a minimum of four hours. If oxygen is available administer it. More severely affected victims may require rescue breathing.

b) Fortunately, the lungs excrete CO within a few hours.

c) Prevention. Ensure there is adequate ventilation when utilizing a fire near your shelter.

6. **MEDICAL AID**. Unfortunately, during a survival situation, a corpsman may not always be available to render assistance. Therefore, the survivor must be knowledgeable in basic first aid as taught in the Marine Battle Skills Training Handbook.

A. Four Life Saving Steps

1) Start the breathing.
2) Stop the bleeding.
3) Treat the wound.
4) Check for shock.

B. **Bites and stings** Reptiles and insects of all kinds can complicate your survival situation. One of the best methods to reducing this risk is to wear your clothing properly. Do not scratch the bite or sting as this can cause the area to become infected.

1) **Bee and Wasp stings**. If stung by a bee, immediately remove the stinger and venom sac by scraping it with a fingernail or knife blade. Wash the sting site thoroughly.

a) Relief for itching and discomfort caused by bites and stings: Cold compress.

b) A cooling paste of mud and ashes.

c) Application of different plants.

2) **Spider bites**. Some people are extremely allergic to the venom associated with spiders. The spiders most venomous in this region are the black widow and the brown recluse. The black widow is black in color with a red hourglass on her belly which has a neurotoxin. The brown recluse has a fiddle on it's back which has a hemotoxin.

3) **Scorpion stings**. Scorpion stings may lead to respiratory distress. There are scorpions in the MWTC area. Scorpions throughout the world generally have neurotoxins.

4) **Snakebites**. There is always the possibility of snakebite throughout regions in the world. Learning the habitats of the indigenous snakes of the area you will be operating in will reduce the chance of being bitten.

a) Deaths from snakebites are rare, and the primary concern in treatment is to limit the amount of eventual tissue destruction around the bite area.

C. **Wounds**. Wounds are actual breaks in the integrity of the skin. Wounds can be caused by accident or by animals. These wounds are most serious in a survival situation, not only because of tissue damage and blood loss, but also because of infection. By taking proper care of the wound you can reduce the chance of a debilitating infection.

D. **Animals. Prevention** of an animal bite is best accomplished through knowledge of behavior, personalities, and patterns.

1) Animals generally give ample warning of their intentions, which are to repel the intruder or permit its escape. Animals that act out of character and approach humans should be considered rabid and avoided.

2) Tearing, cutting, and crushing injuries are combined in animal bites. Always look for secondary injuries.

B. First Aid. Whether the wound was caused by accident or by animal, the treatment remains the same.

1) Early cleansing of the wound reduces the chance of bacterial infection and is extremely effective in removing rabies and other viruses. Cleanse by irrigation. Bleeding wounds also helps the irrigation process initially.

2) Open wound management is best described by the "open treatment" method. Do not try to suture or close the wound. This will seal any dirt or infection into the wound. As long as the wound can drain it will usually not become life threatening.

a) Maggots
b) Super glue
c) Shunt

3) **Tourniquet in a survival situation**. If no rescue or medical aid is likely for over 2 hours, an attempt to **SLOWLY** loosen the tourniquet may be made 20 minutes after it is applied:

a) Ensure pressure dressing is in place.

b) Ensure bleeding has stopped.

c) Loosen tourniquet SLOWLY to restore circulation.

d) Leave loosened tourniquet in position in case bleeding resumes.

e) Bandaging is meant to protect the wound from foreign objects (i.e., dirt).

f) As with any injury to the body you must increase water intake, more so with an open wound.

F. **Herbal Medicines**

1) Consider using herbal medicines only after proper training and when you lack or have limited medical supplies.

2) **WARNING Some herbal medicines are dangerous and may cause further damage or even death.**

7. **CASUALTY EVACUATION**. Casualty evacuation in a cold weather mountainous environment will require a well thought out plan prior to conducting. Poorly planned evacuations will possibly result in additional casualties, lost time, and equipment damage. In a group survival situation expedient litter will have to be constructed in order to transport the patient effectively.

A. General Considerations. **(WSVX.02.15d)** The following considerations are critical for planning a successful evacuation. A useful acronym to use is "APASSNGG".

1) Apply Essential First Aid. (i.e., splints, pressure bandage, etc.)

2) Protect the Patient form the Elements. Provide the casualty with proper insulation and ensure that he is warm and dry.

3) Avoid Unnecessary Handling of the Patient.

4) Select the Easiest Route. Send scouts ahead if possible, to break trails.

5) Set Up Relay Points and Warming Stations. If the route is long and arduous, set up relay points and warming stations to switch stretcher-bearers and assess the casualty.

6) Normal litter teams must be augmented in Arduous Terrain.

7) Give Litter Teams Specific Goals to work towards. This job is extremely tiring, both physically and mentally.

8) Gear. Ensure all of the patient's gear is kept with him throughout the evacuation.

8. **WILDLIFE DISEASES. (WSVX.02.15e)** When handling animals, whether dead or alive, individuals must use preventive measures against possible exposure to wildlife diseases. Although the possibility of disease is remote, certain signs may indicate that an animal may be diseased. The following are some of the more common diseases found in the United States and throughout the world.

A. Hantavirus Pulmonary Syndrome. Hantavirus, or HPS, is a serious respiratory illness that was first recognized in 1993 in an outbreak in New Mexico and Arizona. It is caused by a virus that is carried by a common field rodent called the deer mouse.

1) Method of Transmission. The virus is shed in the droppings, urine, and saliva of the deer mouse. The virus is transmitted to humans when the material dries, becomes airborne and is inhaled.

2) Signs & Symptoms. The disease begins with flu-like symptoms 3 to 45 days after exposure. The disease can rapidly progress into a life-threatening lower respiratory illness characterized by the flooding of the lungs with fluid.

3) Treatment. No cure or vaccine is yet available against infection. The sooner after infection medical treatment is sought, the better the chance of recovery.

4) Prevention. Mice should not be handled; rodent dens should not be disturbed. Package food so that rodents do not crawl all over it. Do not occupy shelters that may have contained rodents.

B. Plague. The cause of plague is *Yersinia pestis*, a bacterium that is maintained in nature through a complex flea-rodent cycle.

1) Method of Transmission. Infection in humans results by flea bites, direct contact with plague-infected rodents, or direct contact with affected non-rodent hosts such as rabbits, hares, cats, and occasionally other animals.

2) Signs & Symptoms. Infection in humans results in severe disease, with a fatality rate of over 50% in untreated cases. An abnormal swelling in the lymph nodes is usually present.

3) Treatment. Infected people must seek medical treatment.

4) Prevention. Animal noted with fleas should be avoided. In a survival situation, a killed animal should be immediately submerged into a cool water source until all the fleas are remove by water or have died.

C. Tick Borne Diseases.

1) Lyme Disease. Infection occurs most often between May and September. In some cases, a characteristic skin rash may develop at the site of the tick bite. The rash may expand to a diameter of 5 inches or more, and there may be an accompanying flu-like illness. If left untreated, infection can lead to chronic disease characterized by neurologic impairment, cardiac problems, or arthritis.

2) Rocky Mountain Spotted Fever. The incubation period in humans is 2-14 days. Initial symptoms are flu-like and commonly include fever, headache, muscle and joint pain, nausea, and vomiting. A rash may appear. The fatality rate in cases that are treated with antibiotics is about 5% and up to 25% that are untreated.

3) Treatment. Infected people must seek medical treatment.

4) Prevention. Daily body inspections should be conducted to remove all ticks.

D. Food Borne Diseases. Food borne illness, frequently called "food poisoning," is acquired by eating food that is contaminated with microbes or their toxins. Live animals may carry the agent, or contamination may occur from another source during processing or preparation of the food.

1) Botulism. Botulism probably is the most widely known and is generally caused by improper storage of meats. Symptoms may begin with vomiting and diarrhea but proceed to the characteristic impaired vision and descending paralysis. Botulism can be fatal.

2) Salmonella. The bacteria are found in the intestinal tracts and feces of a wide range of animals including poultry, swine, cattle, and household pets. Salmonella maybe fatal.

3) Trichinosis. Trichinosis is caused by a parasite contained within the muscle tissue. Most common carriers are wild swine and bear. Thorough cooking of meats will destroy the parasite.

4) Tularemia. Discovered in Tulare County, California. Tularemia has been reported in over 45 species of vertebrates; however, the disease most often involves ticks. It is also commonly found in rabbits and rodents. It is transmittable by uncooked meats or handling contaminated meats with open sores. Tularemia is a life-threatening disease found throughout the world and can only be treated with antibiotics.

5) Treatment. Consuming charcoal will aid in reducing the body's absorption rate. Medical treatment should be sought if available.

6) Prevention. Prevention of food borne diseases can be accomplished by:

   a) Promptly dress game.

   b) Avoid or minimize contamination by gastrointestinal contents.

   c) **Cook food thoroughly.** Internal cooking temperatures should be 165 degrees F or more.

   d) Eat cooked foods immediately.

   e) Store preserved foods properly.

E. Animal Scat. Certain parasites found on scat can infect humans, if the scat is handled unprotected. Raccoon Roundworm can be found in the scat for at least 30 days, while Fox Roundworm will last only approximately 7 days on their scat. Both of these parasites can possibly infect human, which is almost fatal.

REFERENCE:

1. FM 21-76, Survival, 1996.
2. Paul S. Auerbach, Wilderness Medicine, 3rd Edition, 1995.
3. William R. Davidson, Field Manual of Wildlife Diseases in the United States, 2nd Edition, 1997

UNITED STATES MARINE CORPS
Mountain Warfare Training Center
Bridgeport, California 93517-5001

MSVX.02.16
2/6/05

STUDENT HANDOUT

MOUNTAIN WEATHER

TERMINAL LEARNING OBJECTIVE In a survival situation, and given a survival kit, estimate weather conditions, in accordance with the references. **(MSVX.02.16)**

ENABLING LEARNING OBJECTIVES

(1) Without the aid of references, state in writing the cloud progression for both a cold and warm front, in accordance with the references. **(MSVX.02.16a)**

(2) Without the aid of references, describe in writing each type of cloud, in accordance with the references. **(MSVX.02.16b)**

(3) Without the aid of references, state in writing five signs of nature, in accordance with the references. **(MSVX.02.16c)**

(4) Without the aid of references, and an unobstructed view of the sky, state orally the weather forecast for the next twenty four hours, in accordance with the references. **(MSVX.02.16d)**

OUTLINE

1. GENERAL

A. The earth is surrounded by the atmosphere, which is divided into several layers. The world's weather systems are in the troposphere, the lower of these layers. This layer reaches as high as 40,000 feet.

B. Dust and clouds in the atmosphere absorb or bounce back much of the energy that the sun beams down upon the earth. Less than one half of the sun's energy actually warms the earth's surface and lower atmosphere.

C. Warmed air, combined with the spinning (rotation) of the earth, produces winds that spread heat and moisture more evenly around the world. This is very important because the sun heats the Equator much more than the poles and without winds to help restore the balance, much of the earth would be impossible to live on. When the air-cools; clouds, rain, snow, hail, fog and frost may develop.

D. The weather that you find in any place depends on many things, i.e. how hot the air is, how moist the air is, how it is being moved by the wind, and especially, is it being lifted or not?

2. PRESSURE

A. All of these factors are related to air pressure, which is the weight of the atmosphere at any given place. The lower the pressure, the more likely are rain and strong winds.

B. In order to understand this we can say that the air in our atmosphere acts very much like a liquid.

C. Areas with a high level of this liquid would exert more pressure on the Earth and be called a "high pressure area".

D. Areas with a lower level would be called a "low pressure area".

E. In order to equalize the areas of high pressure it would have to push out to the areas of low pressure.

F. The characteristics of these two pressure areas are as follows:

(1) High-pressure area. Flows out to equalize pressure.

(2) Low-pressure area. Flows in to equalize pressure.

G. The air from the high-pressure area is basically just trying to gradually flow out to equalize its pressure with the surrounding air; while the low pressure is beginning to build vertically. Once the low has achieved equal pressure, it can't stop and continues to build vertically; causing turbulence, which results in bad weather.

NOTE: When looking on the weather map, you will notice that these resemble contour lines. They are called "isobars" and are translated to mean, "equal pressure area".

H. Isobars. Pressure is measured in millibars or another more common measurement - "inches mercury".

I. Fitting enough, areas of high pressure are called "ridges" and areas of low pressure are called "troughs".

NOTE: The average air pressure at sea level is:

29.92 inches mercury.
1,013 millibars.

J. As we go up in elevation, the pressure (or weight) of the atmosphere decreases.
EXAMPLE: At 18,000 feet in elevation it would be 500 millibars vice 1,013 millibars at sea level.

3. **HUMIDITY**. Humidity is the amount of moisture in the air. All air holds water vapor, although it is quite invisible.

   A. Air can hold only so much water vapor, but the warmer the air, the more moisture it can hold. When the air has all the water vapor that it can hold, the air is said to be saturated (100% relative humidity).

   B. If the air is then cooled, any excess water vapor condenses; that is, it's molecules join to build the water droplets we can see.

   C. The temperature at which this happens is called the "condensation point". The condensation point varies depending on the amount of water vapor and the temperature of the air.

   D. If the air contains a great deal of water vapor, condensation will form at a temperature of 20°C (68°F). But if the air is rather dry and does not hold much moisture, condensation may not form until the temperature drops to 0°C (32°F) or even below freezing.

   E. Adiabatic Lapse Rate. The adiabatic lapse rate is the rate that air will cool on ascent and warm on descent. The rate also varies depending on the moisture content of the air.

      (1) Saturated Air = 2.2°F per 1,000 feet.

      (2) Dry Air = 5.5°F per 1,000 feet.

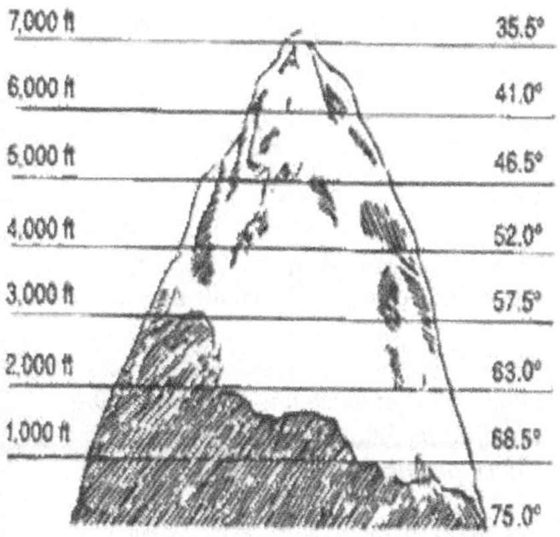

ure 18 **Air temperature rises and cools at an even 5.5 degrees per 1,000 feet when no moisture is present.**

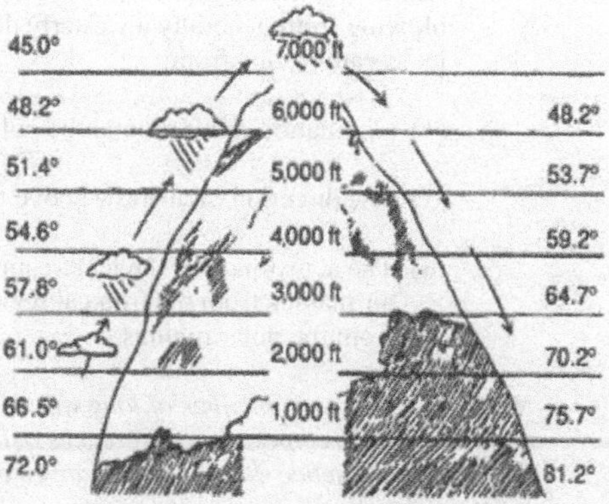

Figure 19 **Add condensation/moisture to the mix and the air cools at a slower rate (approximately 3.2 degrees per 1,000 feet) and then warms at a slower rate too until the moisture evaporates, illustrating adiabatic cooling and warming.**

4. **WINDS**. As we stated earlier, the uneven heating of the air by the sun and rotation of the earth causes winds. Much of the world's weather depends on a system of winds that blow in a set direction. This pattern depends on the different amounts of sun (heat) that the different regions get and also on the rotation of the earth.

A. Above hot surfaces rising air creates a void. Cool air moves into and settles into the void. The cool air is either warmed up and begins to rise or it settles. This is dependent upon the sun's thermal energy. The atmosphere is always trying to equalize between high pressure and low pressure. On a large scale, this forms a circulation of air from the poles along the surface or the earth to the equator, where it rises and moves towards the poles again.

B. Once the rotation of the earth is added to this, the pattern of the circulation becomes confusing.

C. Because of the heating and cooling, along with the rotation of the earth, we have these surfaces winds. All winds are named from the direction they originated from:

(1) Polar Easterlies. These are winds from the polar region moving from the east. This is air that has cooled and settled at the poles.

(2) Prevailing Westerlies. These winds originate from approximately 30 degrees North Latitude from the west. This is an area where prematurely cooled air, due to the earth's rotation, has settled back to the surface.

(3) Northeast Tradewinds. These are winds that originate from approximately 30 degrees North from the Northeast. Also prematurely cooled air.

D. Jet Stream. A jet stream can be defined as a long, meandering current of high speed winds near the tropopause (transition zone between the troposphere and the stratosphere) blowing from generally a westerly direction and often exceeding 250 miles per hour. The jet stream results from:

(1) Circulation of air around the poles and Equator.

(2) The direction of air flow above the mid latitudes.

(3) The actual path of the jet stream comes from the west, dipping down and picking up air masses from the tropical regions and going north and bringing down air masses from the polar regions.

NOTE: The average number of long waves in the jet stream is between three and five depending on the season. Temperature differences between polar and tropical regions influence this. The long waves influence day to week changes in the weather; there are also short waves that influence hourly changes in the weather.

E. Here are some other types of winds that are peculiar to mountain environments but don't necessarily affect the weather:

(1) Anabatic wind. These are winds that blow up mountain valleys to replace warm rising air and are usually light winds.

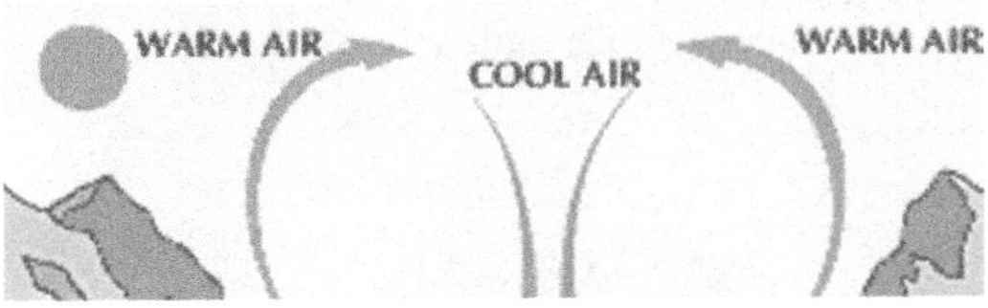

(2) Katabatic wind. These are winds that blow down mountain valley slopes caused by the cooling of air and are occasionally strong winds.

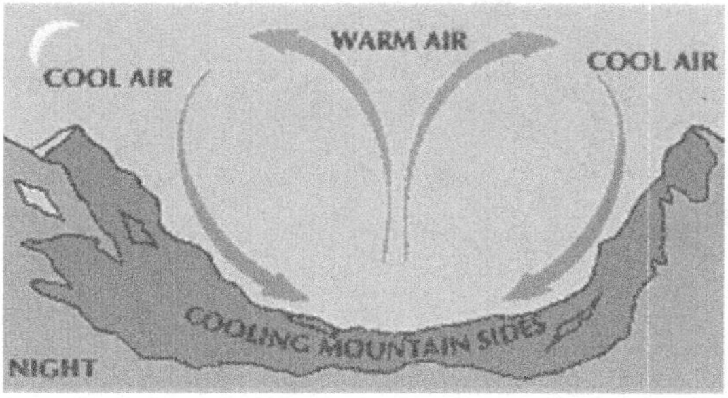

5. **AIR MASSES**. As we know, all of these patterns move air. This air comes in parcels known as "air masses". These air masses can vary in size from as small as a town to as large as a country. These air masses are named for where they originate:

A. Maritime. Over water.

B. Continental. Over land.

C. Polar. Above 60 degrees North.
D. Tropical. Below 60 degrees North.

E. Combining these give us the names and description of the four types of air masses:

(1) Continental Polar. Cold, dry air mass.

(2) Maritime Polar. Cold, wet air mass.

(3) Continental Tropical. Dry, warm air mass.

(4) Maritime Tropical. Wet, warm air mass.

F. The thing to understand about air masses, they will not mix with another air mass of a different temperature and moisture content. When two different air masses collide, we have a front which will be covered in more detail later in this period of instruction.

6. **LIFTING/COOLING**. As we know, air can only hold so much moisture depending on it's temperature. If we cool this air beyond its saturation point, it must release this moisture in one form or another, i.e. rain, snow, fog, dew, etc. There are three ways that air can be lifted and cooled beyond its saturation point.

A. Orographic uplift. This happens when an air mass is pushed up and over a mass of higher ground such as a mountain. Due to the adiabatic lapse rate, the air is cooled with altitude and if it reaches its saturation point we will receive precipitation.

OROGRAPHIC UPLIFT

B. Convention effects. This is normally a summer effect due to the sun's heat radiating off of the surface and causing the air currents to push straight up and lift air to a point of saturation.

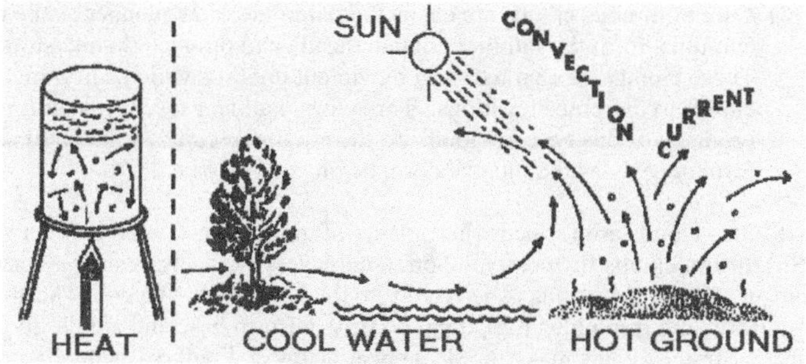

CONVECTION EFFECTS

C. Frontal lifting. As we know when two air masses of different moisture and temperature content collide, we have a front. Since the air masses will not mix, the warmer air is forced aloft, from there it is cooled and then reaches its saturation point. Frontal lifting is where we receive the majority of our precipitation. A combination of the different types of lifting is not uncommon.

7. **CLOUDS**. Anytime air is lifted or cooled beyond its saturation point (100% relative humidity), clouds are formed. Clouds are one of our sign posts to what is happening. Clouds can be described in many different ways, they can also be classified by height or appearance, or even by the amount of area covered, vertically or horizontally. **(MSVX.02.16b)**

A. Cirrus. These clouds are formed of ice crystals at very high altitudes (usually 20,000 to 35,000 feet) in the mid-latitudes and are thin, feathery type clouds. These clouds can give you up to 24 hours warning of approaching bad weather, hundreds of miles in advance of a warm front. Frail, scattered types, such as "mare-tails" or dense cirrus layers, tufts are a sign of fair weather but predictive may be a prelude to approaching lower clouds, the arrival of precipitation and the front.

B. Cumulus. These clouds are formed due to rising air currents and are prevalent in unstable air that favors vertical development. These currents of air create cumiliform clouds that give them a piled or bunched up appearance, looking similar to cotton balls. Within the cumulus family there are three different types to help us to forecast the weather:

(1) Cotton puffs of cumulus are Fair Weather Clouds but should be observed for possible growth into towering cumulus and cumulonimbus.

(2) Towering cumulus are characterized by vertical development. Their vertical lifting is caused by some type of lifting action, such as convective currents found on hot summer afternoons or when wind is forced to rise up the slope of a mountain or possibly the lifting action that may be present in a frontal system. The towering cumulus has a puffy and "cauliflower-shaped" appearance.

(3) Cumulonimbus clouds are characterized in the same manner as the towering cumulus, form the familiar "thunderhead" and produce thunderstorm activity. These clouds are characterized by violent updrafts which carry the tops of the clouds to extreme elevations. Tornadoes, hail and severe rainstorms are all products of this type of cloud. At the top of the cloud, a flat anvil shaped form appears as the thunderstorm begins to dissipate.

C. Stratus. Stratus clouds are formed when a layer of moist air is cooled below its saturation point. Stratiform, clouds lie mostly in horizontal layers or sheets, resisting vertical development. The word stratus is derived from the Latin word "layer". The stratus cloud is quite uniform and resembles fog. It has a fairly uniform base and a dull, gray appearance. Stratus clouds make the sky appear heavy and will occasionally produce fine drizzle or very light snow with fog. However, because there is little or no vertical movement in the stratus clouds, they usually do not produce precipitation in the form of heavy rain or snow.

7. **FRONTS**. As we know, fronts often happen when two air masses of different moisture and temperature content interact. One of the ways we can identify that this is happening is by the progression of the clouds. **(MSVX.02.16a)**

A. Warm Front. A warm front occurs when warm air moves into and over a slower (or stationary) cold air mass. Since warm air is less dense, it will rise naturally so that it will push the cooler air down and rise above it. The cloud you will see at this stage is cirrus. From the point where it actually starts rising, you will see stratus. As it continues to rise, this warm air cools by the cold air and, this, receiving moisture at the same time. As it builds in moisture, it darkens becoming "nimbus-stratus", which means rain of thunderclouds. At that point some type of moisture will generally fall.

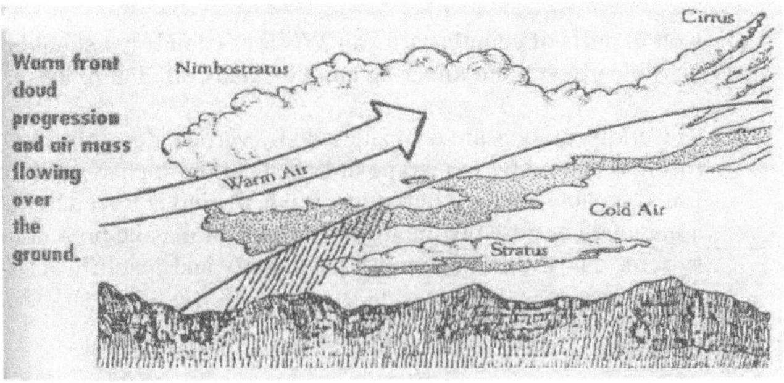

B. Cold Front. A cold front occurs when a cold air mass (colder than the ground that it is traveling over) overtakes a warm air mass that is stationary or moving slowly. This cold air, being denser, will go underneath the warm air, pushing it higher. Of course, no one can see this, but they can see clouds and the clouds themselves can tell us what is happening. The cloud progression to look for is cirrus to cirrocumulus to cumulus and, finally, to cumulonimbus.

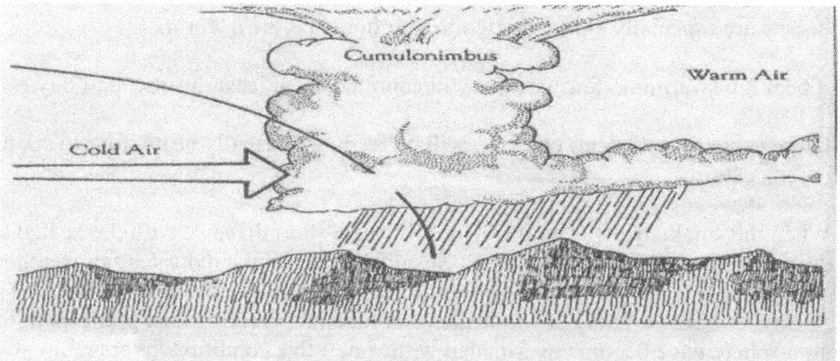

Cold front cloud progression and air mass flowing over the ground.

C. Occluded Front. Cold fronts move faster than warm ones so that eventually a cold front overtakes a warm one and the warm air becomes progressively lifted from the surface. The zone of division between cold air ahead and cold air behind is called a "cold occlusion". If the air behind the front is warmer than ahead, it is a warm occlusion. Most land areas experience more
occlusions than other types of fronts. In the progression of clouds leading to fronts, orographic uplift can play part in deceiving you of the actual type of front, i.e. progression of clouds leading to a warm front with orographic cumulus clouds added to these. The progression of clouds in an occlusion is a combination of both progressions from a warm and cold front.

8. **USING SIGNS FROM NATURE**. **(MSVX.02.16c)** These signs will give you a general prediction of the incoming weather conditions. Try to utilize as many signs together as possible, which will improve your prediction. All of these signs have been tested with relative accuracy, but shouldn't be depended on 100%. But in any case you will be right more times than wrong in predicting the weather. From this we can gather as much information as needed and compile it along with our own experience of the area we are working in to help us form a prediction of incoming weather. The signs are as follows:

A. Contrail Lines. A basic way of identifying a low-pressure area is to note the contrail lines from jet aircraft. If they don't dissipate within two hours, that indicates a low pressure area in your area. This usually occurs about 24 hours prior to an oncoming front.

B. Lenticulars. These are optical, lens-shaped cumulus clouds that have been sculpted by the winds. This indicates moisture in the air and high winds aloft. When preceding a cold front, winds and clouds will begin to lower.

C. An altimeter and map or a barometer can be utilized to forecast weather in the field. However, the user must have operational knowledge of the gear.

D. A spider's habits are very good indicators of what weather conditions will be within the next few hours. When the day is to be fair and relatively windless, they will spin long filaments over which they scout persistently. When precipitation is imminent, they shorten and tighten their snares and drowse dully in their centers.

E. Insects are especially annoying two to four hours before a storm.

F. If bees are swarming, fair weather will continue for at least the next half day.

G. Large game such as deer, elk, etc., will be feeding unusually heavy four to six hours before a storm.

H. When the smoke from a campfire, after lifting a short distance with the heated air, beats downward, a storm is approaching. Steadily rising smoke indicates fair weather.

I. A gray, overcast evening sky indicates that moisture carrying dust particles in the atmosphere have become overloaded with water; this condition favors rain.

J. A gray morning sky indicates dry air above the haze caused by the collecting of moisture on the dust in the lower atmosphere; you can reasonably expect a fair day.

K. When the setting sun shows a green tint at the top as it sinks behind clear horizon, fair weather is probable for most of the next 24 hours.

L. A rainbow in the late afternoon indicates fair weather ahead. However, a rainbow in the morning is a sign of prolonged bad weather.

M. A corona is the circle that appears around the sun or the moon. When this circle grows larger and larger, it indicates that the drops of water in the atmosphere are evaporating and that the weather will probably be clear. When this circle shrinks by the hour, it indicates that the water drops in the atmosphere are becoming larger, forming into clouds, rain is almost sure to fall.

N. In the northern hemisphere winds form the south usually indicate a low-pressure system. These systems are frequently associated with rainstorms. "Winds from the south bring s rain in it's mouth."

O. It is so quiet before a storm, that distant noises can be heard more clearly. This is due to the inactivity of wildlife a couple of hours before a storm.

P. Natural springs tend to flow at a higher rate when a storm is approaching. This is due to lower barometric pressure. This will cause ponds, with a lot of vegetative decay at the bottom, to become momentarily polluted.

Q. A heavy dew or frost in the morning is a sign of fair weather for the rest of the day. This is due to the moisture in the atmosphere settling on the ground vice in the form of precipitation and up to 12 hours of continued good weather can be expected.

REFERENCE:

1. William J. Kotsch, Rear Admiral, USN RET., Weather For the Mariner, 3rd Edition

UNITED STATES MARINE CORPS
Mountain Warfare Training Center
Bridgeport, California 93517-5001

MSVX.02.17
2/6/05

STUDENT HANDOUT

INTRODUCTION TO EVASION

TERMINAL LEARNING OBJECTIVE In a survival situation, and given a survival kit, employ evasion techniques, in accordance with the references. **(MSVX.02.17)**

ENABLING LEARNING OBJECTIVES

(1) Without the aid of reference, list in writing the planning and preparation considerations for evasion, in accordance with the references. **(MSVX.02.17a)**

(2) Without the aid of reference, describe in writing the definition of a Selected Area For Evasion (SAFE), in accordance with the references. **(MSVX.02.17b)**

(3) Without the aid of reference, list in writing the steps taken during the occupation of a SAFE, in accordance with the references. **(MSVX.02.17c)**

(4) Without the aid of reference, conduct basic evasion techniques, in accordance with the references. **(MSVX.02.17d)**

OUTLINE

1. **PREPARING FOR A POTENTIAL EVASION SITUATION.** The Code of Conduct provides guiding principles to Marines involved in any military operation whether peacekeeping, combat, or survival. An operation that deteriorates so severely that a Marine unit is forced to employ survival skills may require that unit to "evade" hostile enemy units. JP 3-50.3 defines evasion as the process whereby individuals who are isolated in hostile or unfriendly territory avoid capture with the goal of successfully returning to areas under friendly control. Should a survival situation require evading the enemy, success will depend on prior planning.

   a. Planning and Preparation. **(MSVX.2.16a)** The responsibility for proper preparation and planning for evasion ultimately rests with the individuals concerned. All Marines who are tasked to execute any mission should receive the following:

      (1) Intelligence Briefings. Information on the mission route, enemy troop dispositions, impact of enemy operations on friendly or multinational military forces, status of the US or multinational military situation, or changing attitudes of the enemy populace.

(2) Evasion Plan of Action (EPA). The EPA is one of the critical documents for successful recovery planning. It is the vehicle by which potential evaders, prior to their isolation in hostile territory, relay their after-isolation intentions to the recovery forces. See Appendix D, "Evasion Plan of Action Format," for details on the content of an EPA.

(3) Selected Areas for Evasion (SAFE) Area Intelligence Descriptions. **(MSVX.2.16b)** A SAFE is a "designated area in hostile territory that offers isolated personnel a reasonable chance of avoiding capture and of surviving until they can be recovered."

(a) They are designated by the Defense Intelligence Agency (DIA) and are classified.

(b) Designed to facilitate extended evasion, which must meet certain requirements for approval.

(4) E&R (Evasion and Recovery) Area Studies. E&R areas may be selected in any geographic region based on operational or contingency planning requirements. Although similar to SAFE areas in most respects, they differ in that not all conventional selection criteria for SAFE areas can be met because of current political, military, or environmental factors prevailing in the country.

(5) Survival, Evasion, Resistance, and Escape Guides and Bulletins. They contain the basic information to help an individual survive, successfully evade and, if captured, resist enemy exploitation. These bulletins cover information on topography and hydrography, food and water sources, safe and dangerous plants and animals, customs and cultures.

(6) Isolated Personnel Report (ISOPREP). When filled in, the DD Form 1833 is classified CONFIDENTIAL. It enables a recovery force to authenticate evaders.

2. **EXECUTING AN EVASION PLAN OF ACTION (EPA).** Unforeseen circumstances may require Marines to execute their EPA.

a. Initial Planning. Immediately upon breaking contact, attempt to gain maximum distance between yourself and the enemy.

(1) Carefully consider METT-T during all planning and execution.

(2) Determine unit's combat effectiveness.

(3) Develop a course of action.

b. Movement techniques. If possible, the entire movement to friendly or neutral areas, as well as to designated SAFE areas or E&R areas should be completed without being observed. Furthermore, an appreciation of the methods by which a hostile force may attempt to detect you will assist in techniques to maximize your concealment.

(1) Methods to avoid enemy detection.

(a) Apply standard patrolling movement techniques.

(b) Avoid natural lines of drift and Main Supply Routes (MSR).

(c) Avoid all rural areas, small towns, and farms.

-Dogs and domestic poultry are very common and will provide a "first alert" needed to initiate a hostile search.

(2) Methods of detecting the evader.

(a) Direct Observation.

(b) Detection Equipment.

-Thermal imaging
-Active Infrared (IR), such as NVGs
-Acoustic detectors/sensors
-Direction finding equipment for radios

(c) Search teams.

-Military and/or civilian
-Trackers

(d) Dogs.

-Attack or tracking dogs
-Difficult to determine if being tracked by dogs
-Attempt to discourage the dog from doing its job

c. Occupation of a SAFE or E&R. **(MSVX.2.16c)** Prior to movement to, and occupation of a SAFE or E&R area, consider the following:

(1) Conduct a reconnaissance of the entire area for enemy threat. This may me a physical or visual reconnaissance.

(2) Select an occupation site which affords:

(a) Concealed escape routes if detected by enemy.

(b) Close proximity to a potential extraction site.

(c) Observation of the area and avenues of approach.

(3) Apply the requirements for survival.

(4) Execute the communication and signaling plan as ordered.

<u>REFERENCE</u>.

1. JP 3-50.3, <u>Joint Doctrine for Evasion and Recovery</u>, 1996.

UNITED STATES MARINE CORPS
Mountain Warfare Training Center
Bridgeport, California 93517-5001

EVASION PLAN OF ACTION FORMAT

1. Individuals completing EPAs should not use the statement "PER SAR SPINS (Special Instructions) as a substitute for this document. Such a statement fails to provide recovery forces with the information required and provides no concrete data with which to plan a recovery operation.

2. EPAs should contain the following minimum information. Inclusion of this prescribed information into one document enhances operational effectiveness and precludes the possibility that critical information might not be available in a time-sensitive situation. These documents must be classified to at least the level of the operation order for the mission they support. Paragraphs must be individually classified to the appropriate level.

   a. Indentification

      (1) Name and rank of each individual.

      (2) Mission number, aircraft, or call sign.

   b. Planned Route of Flight or Travel.

      (1) Route points must be describe in the EPA for both ingress and egress.

      (2) Describe inflight emergency plans for each leg of the mission.

   c. Immediate Evasion Actions and/or Intentions for the First 48 hours, Uninjured (for example):

      (1) Hide near aircraft site or area of separation from unit (distance and heading).

      (2) Evade alone or link-up with others at rally point.

      (3) Travel plans (distance, duration or time, speed, and other such details).

      (4) Intended actions and/or length of stay at initial hiding location.

   d. Immediate Evasion Actions and/or Intentions, If Injured.

      (1) Provide hiding intentions if injured.

      (2) Provide evasion intentions if injured.

      (3) Provide travel intentions if injured.

(4) Provide intended actions at hiding locations if injured.

e. Extended Evasion Actions and/or Intentions After 48 hours.

(1) Destination (SAFE, mountain range, coast, border, FEBA).

(2) Travel routes, plans, and/or techniques (either written and/or sketched).

(3) Actions and/or intentions at potential contact or recovery locations.

(4) Recovery/contact point signals, signs, and/or procedures (written out and/or sketched).

(5) Back-up plans, if any, for the above.

3. The following information should be completed by appropriate communications and/or signal, intelligence personnel and attached to the EPA.

a. Communications and Authentication.

(1) Codewords.

(2) Available communications and signaling devices.

(3) Primary communication schedule, procedures, and/or frequencies (first 48 hours and after 48 hours).

(4) Back-up communication schedule, procedures, and/or frequencies.

b. In addition to the above minimum required information, units may wish to include the following optional information:

(1) Weapons and ammunition carried.

(2) Personal evasion kit items.

(3) Listing of issue survival and evasion kit items.

(4) Mission evasion preparation checklist

(5) Signature of reviewing official.

PME VIDEO GUIDED DISCUSSION

"THE EDGE"

BREAK AT THE FLIGHT TO FIND THE INDIAN

1. Discuss the bear threat, emphasize bivouac routine and food storage.
2. Discuss amount of available sunlight in northern latitudes (winter/summer).
3. Discuss book knowledge vs. skills.
4. Explain the Taiga Ecosystem and where it ranges in the world.

BREAK AFTER THE FIRST DAY

1. Discuss their "mind-set/attitude" at the cabin - childish & unprepared with no survival kit.
2. What should their first concern be after the cold water immersion?
3. "Any one got matches?" - How did they stay dry?
4. What tinder did they have for their fire? Witches hair should have been used.
5. What priorities of work did they accomplish?

BREAK AT THE NIGHT FIRE

1. "Die of shame", what natural reaction to stress does it relate to?
2. Watch method/improvised compass - problems for survival navigation.
3. Why was it a good decision to travel?
4. What should they have done prior to traveling?
5. Group survival - Point out how as a group, the weak became strong when they formulated a plan together & how the weak became strong when tasked.

BREAK AT THE HELO FLYING OVER

1. Discuss the thatching job on their shelter.

2. Discuss commitment/courage about the bloody bandages. "What difference does it make?"

3. Discuss snow storms in mountains at any time.

4. Discuss survival signaling - be prepared.

BREAK AFTER THE BEAR KILL

1. Discuss how they passed up enormous food sources.
   -Reindeer moss
   -Fish

2. Discuss how they should be thinking of long term survival prior to the onset of winter.

3. Point out how bears run straight through the woods.

4. Point out when a bear will charge: pawing back & forth with head swaying side to side.

5. Point out that man will not out run a bear.

6. Explain that a mauled survivor was only disciplined by the bear.

DISCUSS AT THE END OF THE MOVIE

1. How tools became important.

2. The feasibility of digging large pit traps.

3. How Anthony Hopkins sat & thought prior to making a decision - "Undue haste makes waste".

4. How realistic the amount of bough would create the amount of smoke generated.

5. Discuss finally how knowledge comes first, but knowledge without skills is useless!

SURVIVAL QUICK REFERENCE CHECKLIST

S - **S**ize up the situation, surroundings, physical condition, & equipment.
U - **U**ndue haste makes waste.
R - **R**emember where you are.
V - **V**anquish fear and panic.
I - **I**mprovise & Improve.
V - **V**alue living.
A - **A**ct like the natives.
L - **L**ive by your wits, *but for now*, Learn Basic Skills.

1. IMMEDIATE ACTIONS

a. Assess immediate situation...**THINK BEFORE YOU ACT!**
b. Take action to protect yourself from NBC hazards.
c. Seek concealment.
d. Assess medical condition; treat as necessary.
e. Sanitize uniform of potentially compromising information.
f. Sanitize area, hide equipment you are leaving.
g. Apply camouflage.
h. Move away from initial site using patrolling techniques.
I. Use terrain to advantage; cover, concealment, and communication advantage.
j. Find a rally point with the following:
 -Cover and concealment.
 -Safe distance from enemy positions and Lines of Communication (LOCs).
 -Multiple avenues of concealed escape routes.
 -Has locations for LPs and OPs.
 -Protection from the elements.
 -Near a reliable water and fuel source.
 -Location for possible communication/signaling opportunities.

2. RALLY POINT

a. Establish security: treat injuries, inventory equipment, improve camouflage.
b. Assess Commanders Intent IAW ability to execute mission.
c. Determine level of combat effectiveness.
d. Develop a course of action using METT-TSL; establish priorities.
e. Execute course of action...stay flexible!

3. MOVEMENT

a. Travel slowly and deliberately.
b. Do not leave evidence of travel, use noise and light discipline.
c. Stay away from LOCs.
d. Use standard patrolling techniques.

4. CHANCE CONTACT

a. Only engage the enemy with reasonable chance of success.
b. Use METT-TSL for all engagements.
c. Break contact for all unfavorable engagements.

5. COMMUNICATION AND SIGNALING

a. Communicate per theater communication procedures, particularly when considering transmitting in the "blind".
b. Be prepared to use signaling devices on short notice.
c. Execute signaling per mission order.

6. RECOVERY OPERATIONS

a. Select site(s) IAW mission order.
b. Ensure site is free of hazards and enemy.
c. Select best area for communications and signaling devices.
d. Observe site for proximity to enemy activity and LOCs.
e. Follow recovery force instructions.

ANIMAL HABITS

(1) Coyote and Wolf. Coyotes and wolves often run in family groups, especially in the early part of the season. When a littermate is caught, normally other coyotes will return to the set site, so reset traps in the same area. They are inquisitive, so you want them to smell and see. Generally they run 3-25 square mile territories, even larger during periods of rough weather. They will move and congregate around a good food source until it has been eaten. Some years, natural food abundance will have them working one species and showing little interest for other foods and baits. Look at fresh scat and select baits on what they are eating. Coyotes are found at every elevation and habitat type in North America while wolves are restricted to northwestern states. At higher elevations during deep winter snows, coyotes will move to lower elevations with the deer, elk, and livestock, although some will tough it out in the deep snow.

(2) Fox. Gray, Red, Prairie Swift, and Desert Kit foxes are found throughout the U.S. Grays are found in pinion-juniper, cedar, oak brush, canyon bottoms and hogbacks, cottonwood draws and edges where these meet. The Red Fox is found in irrigated agricultural lands along the bases of mountain ranges and prairie rivers, in the high mountain parks and alpine. The Prairie Swift Fox is found in the prairie states of this country. The desert kit fox is found in the southwestern corner states. Both of these species are vary curious. Fox habits are very much like the coyote.

(3) Bobcat and Lynx. Bobcats are generally found in the west while Lynx are found in the northwestern states. They compete with coyotes and sometimes are preyed on by coyotes. Their territory is generally two square miles. They den in rock caves, deadfalls, hollow trees and logs. They are sight hunters and use their eyes and ears more than the sense of smell. They prefer to kill their own food and avoid rotten carrion. Bobcat can be "pulled" to an area by curiosity lures. They often avoid large open space.

(4) Raccoon and Opossum. They are located throughout the U.S. They like a combination of water, old mature trees, buildings and junk piles, and a consistent food supply like grain or prepared feed.

(5) Ring-tailed Cat. They inhabit watercourses where rocky canyons or broken rock, erosion holes and rough terrain occurs. They are found in the western states. Old timers talk of ringtails being fairly abundant prior to the expansion of the raccoon range and densities during and after WWII. It is possible that raccoons have replaced the ringtail in much of its former habitat.

(6) Badger. They are found from above the timberline to the lowest elevation in the west. They apparently do not tolerate high densities and generally there are only 3-4 per square mile while running 1-3 square miles. They prefer rodents but take carrion, fruit, insects, roots and grain. They hole up for long periods during extremely cold weather, moving just before and after severe weather systems. These animals can be tracked to dens and snared in the den. They often inhabit prairie dog towns.

(7) Skunk. Three species are found in the west: striped, spotted, and hognose. Skunks are located just about everywhere. Their musk is a prized ingredient for lures.

(8) Weasel. Two species of weasels occur in most of the west: the ermine or short-tailed, and the long-tailed. They prefer meat and blood, although sometimes they are caught on peanut butter. Their body shape is adapted for living and pursuing their prey on the prey's own territory, burrows, tunnels, and runways. They have ferocious appetites and will tackle grouse, rabbits, ducks, and squirrels. They are inquisitive and can't pass up examining cavities, knot holes, and burrow entrances.

(9) Marten. Marten are found in the upper montane and subalpine zones above 8,000 feet. They generally don't venture far from the escape cover of trees. They live on squirrels, rabbits, voles, deer mice, grouse and other small birds and mammals. There may be as many as 3-5 per square mile. They like fresh kidney, heart, liver, and spleen for baits.

(10) Mink. They are found in suitable water habitat or marshy ground with good bank development and undisturbed wetlands vegetation. They feed on fish, crustaceans, clams, and small mammals and birds. They will general travel along the stream bank.

(11) Beaver. They are found in almost all water with cottonwood, aspen, or willow trees. Peak activity of beaver is from September to freeze-up when they are repairing dams, lodges and building food caches. There will usually be slides from the water to cutting areas of trees. Their castor is an excellent additive to lure.

(12) Muskrat. They are found in springs, dugouts, dams, and permanent pools on intermittent streams, rivers, and irrigation ditches, mountain lakes and beaver ponds. Their dens are usually below the water line and into the bank.

(13) Black Bear. They are found throughout North America. The female generally has one or two cubs during the winter hibernation. Survival of the cubs is good since they benefit from at least a year of parental

care. Sows have litters every other year and will not produce cubs until 2 or 3 years of age. They are omnivorous in their feeding, taking what is available such as insects, rodents, berries, roots, fish, and carrion. They are inquisitive. Their color can range from black, brown, cinnamon, to golden.

(14)Mountain Lion. In colonial America the mountain lion, painter, panther, catamount, or cougar was found throughout the U.S. Much of their disappearance was due to the clearing of forests and land-use changes as development progressed. They have litters of 1-6 kittens, averaging 2-3 normally. These are generally born in late winter and early spring. Dens are generally caves in rocky country, hollow logs, windfall trees and various cavities that provide protection from weather. Males can weigh up to 276 pounds and 175 pounds for females. They have tremendous strength with reports of mature lions carrying full-grown deer up cliffs, moving 650-pound cattle, and carrying adult elk for long distances. They prefer to kill their own food and disdain soured and decaying carrion. Kills are deep scratches and gouges on the neck and shoulders, bites and scratches around the neck, face and eyes. The skin is peeled back and the blood rich liver, spleen, kidneys, and lungs are eaten first. The muscle tissue is gnawed from the bones. The lion will normally scratch dirt and plant material over the kill, leave it and return for a future meal.

(15)Rabbits and Hares. There are numerous species located throughout North America with the Cottontail, Black-Tailed Jack, and the white-tailed Jack being located in our training area. During winter months, they will feed on aspen and willow twigs.

UNITED STATES MARINE CORPS
Mountain Warfare Training Center
Bridgeport, California 93517-5001

TACTICAL CONSIDERATIONS

If the need arises to implement survival skills in a semi-permissive or non-permissive environment the Marine must be able to utilize basic skills, as referenced in the Marine Battle Skills Handbook Pvt – Lcpl, in order to avoid making contact with hostile personnel. Unfortunately, the enemy will not consider your **MOS** when deciding whether or not you should be **captured** or **killed**. It is imperative that every Marine live up to statement – **"every Marine is a basic rifleman."**

To discuss every possible survival scenario and enemy situation would be pointless. The following outline is to be utilized as a guide. Common sense and survival skills, along with these considerations, will increase your chances of avoiding capture or possible death.

A. Apply the key word survival (ICBT – 20.01) (PVTX.14.16)

1. **Size up the situation:**
   - **Mission**
     What was the mission?
     Can the mission still be accomplished?
   - **Enemy**
     What is the enemy situation?
   - **Troops and Fire Support available**
     Do you have communications with higher?
     Is anyone injured? How will they be transported?
     How will you communicate with other Marines in your group?
     Arm and Hand signals?
   - **Terrain and Weather**
     Do you have a map?
     Does the terrain offer cover and concealment?
     Where is the water?
     Do you have protection from the elements?
     Is it advantageous to move in current weather?
     Will you move in daytime or evening temperatures?
     How much illumination is available at night?
   - **Time, Space, and Logistics**
     Is it day or night?
     Time Distance Formula
     What kind of supplies and equipment are available?

2. **Undue haste makes waste:**
   - Should you stay or move from your current position.

- "Slow is Smooth – Smooth is Fast" i.e. Is there a need to run to the SAFE or should the requirements of survival be implemented in route? Security is paramount -is it being sacrificed for speed?

3. **Remember where you are:**
   - Are you in a non-permissive environment?
   - What is the terrain like?
   - Can you utilize land navigation skills?

4. **Vanquish fear and panic:**
   - Are good decisions being made?
   - Is the group completely lost and leaderless?
   - **BAMCIS**

5. **Improvise and improve:**
   - Do you have your survival kit?
   - Are litters available or do you have to improvise?
   - Do you have the resources to obtain food and water?
   - Will your supplies and equipment protect you from the elements?
   - Will your supplies and equipment protect you from the enemy?

6. **Value living:**
   - Do you want to lay on your back and put your legs in the air like a dead cockroach?

7. **Act like the natives:**
   - Observe native habits.

8. **Live by your wits, but for now learn basic skills:**
   - Utilize common sense and basic Marine Corps training.
   - Practice skills learned at MWTC.
   - **Prior planning prevents poor performance.**
   - Establish E and R plan, brief personnel of contingencies.

B. **Additional Individual Training Standards**:
- Employ signaling devices **(ICBT – 20.03)**
- Construct and maintain a fire **(ICBT – 20.04)**
  - What is the tactical situation?
- Prepare a survival kit **(ICBT 20.05) (PVTX.14.15)**
- Maintain the M16A2 service rifle. **(PVTX.11.1)**
- Prepare individual equipment for tactical operations. **(PVTX.14.1)**
- Camouflage self and individual equipment. **(PVTX.14.7)**
- Transport casualties using manual carries and improvised stretchers. **(PVTX.17.4)**
- Maintain physical fitness. **(PVTX.20.1)**
- Perform individual movement. **(PVTX.14.2)**
- Participate in a security patrol. **(PVTX.13.1)**

        - Arm and hand signals.
        - Challenge and pass / near and far recognition.
        - Rally points.
        - Actions on enemy contact
    - Security halts.
    - Reconnaissance of objectives.
    - 5 Point Contingency Plan. (GOTWA)
        - Going (where)
        - Others (who is going with you)
        - Time away.
        - What happens (you and them)
        - Actions taken on enemy contact (you and them)
- React to enemy indirect fire. **(PVTX.14.3)**
- React to enemy direct fire. **(PVTX.20.1)**

C. Additional Considerations:

When sizing up the situation you will determine whether the Mission or Cmdr's Intent can be accomplished. Obviously, every effort must be made to accomplish the mission. If it can not be accomplished a separate mission order must be established for the group. (i.e. At 0900 the group will implement the requirements for survival in order to move to the SAFE for recovery.)

In order to tactically move and occupy the SAFE, the group must be task organized into teams. The teams are assigned additional tasks. Team tasks will usually be accomplished during occupation of the patrol base. In addition, individuals within the group are assigned tasks.

1. Teams:

- Security. All patrol members should be assigned sectors of fire to include air sentry.
- Reconnaissance.
- Assault. (may not be implemented)
- Support. (may not be implemented)

2. Team Tasks:

- Water procurement.
- Food gathering.
- Wood gathering. (construction materials, signal, and fire wood.)
- Shelter construction.
- Pathguards.
- Signaling

3. Individual Tasks:

- Point man.
- Navigator.
- Patrol leader and assistant.

- Flank (left and right) security.
- Two pace men.
- Tail end charlie.

In a survival situation it is probably wiser to occupy the patrol base through reconnaissance instead of by force. Upon initial occupation the acronym SAFE (Security, Automatic weapons, Fields of fire, Entrenchment) must be enforced. The patrol base should only be entered (or exited) from one location. Communication within the group is essential. Everyone must be well informed. (i.e. current plans, alternate patrol bases or rally points, how many Marines have departed the patrol base and when are they expected to return, current challenge and pass, and near and far recognition signals.) No Marine will exercise their judgement and leave the patrol base or perform a task without permission from the patrol leader.

The patrol base is not a place for lollygagging. It is a place where noise and light discipline is enforced and security is maintained continuously. Priorities of work will be established after occupation. (i.e. weapon maintenance, hygiene, chow, and rest plans) Security or reconnaissance teams can be sent out to determine enemy threat or gather information for route selection. These patrols can also be tasked with gathering firewood or some other routine task. However, security must be maintained while the firewood is being collected.

Since you are probably evading the enemy, activity in and around the patrol base must be limited. Occupation of the patrol base must not exceed 24 hours. Depending on the tactical situation the use of fire may or may not be appropriate.

D. Conclusion:

The above information servers as a guideline. Survival is a thinking person's challenge between life and death. As the situation changes a Marine must adapt, size up the situation, and implement a new plan in order to survive. If you are lacking in any of the Individual Training Standards it is your responsible to take corrective action. **Remember that every Marine is a basic rifleman.**

UNITED STATES MARINE CORPS
Mountain Warfare Training Center
Bridgeport, California 93517-5001

WINTER MOUNTAIN SURVIVAL COURSE

PERFORMANCE EVALUATION RECAP SHEET

1. The student must satisfactorily achieve an 80% or higher grade on either the written test or the written retest.

2. The student must satisfactorily master 10 out of the 12 performance evaluation tasks.

3. A breach of integrity, conduct unbecoming of a student or the inability to abide by Survival course guidelines, will result in the immediate dismissal from the Mountain Survival Course.

4. All of the above must be accomplished prior to successful graduation from the Mountain Survival Course.

5. Failure to accomplish any one these will result in a Proficiency Board, which may result in dismissal from the course.

6. There will be no talking to any student during isolation unless an emergency. No student will come with in 50 meters of another student.

PERFORMANCE TASK	M--N/M	REMARKS

All criteria listed below each task, must be accomplished in order to master those tasks.

1. SURVIVAL KIT (MSV.2.2) M / NM
   Fire starting items
   Water procurement items
   Food procurement items
   Signaling items
   First Aid items
   Shelter items

2. SURVIVAL NAVIGATION (MSV.2.7) M / NM
   Can find cardinal directions
   Prepares and maintains a log book
   Uses steering marks
   Appropriate place

3. BOW & DRILL (MSV.2.13a) M / NM
   Bow
   Drill
   Socket
   Fire Board
   Ember Patch
   Birds Nest
   Kindling
   Fuel Wood

4. SURVIVAL SHELTERS (MSV.2.4) M / NM
   Protection from the elements
   Heat retention
   Ventilation
   Drying facility
   Free from hazards
   Shelter stable

5. REQUIREMENTS FOR SURVIVAL (MSV.2.1) M / NM
   1st 24Hrs.
   Shelter
   Fire
   Water
   Signaling

   2nd 24Hrs.
   Tools & Weapons
   Traps & Snares
   Path guards

6. IMPROVISED SIGNAL DEVICE (MSVX.2.6) M / NM
   Smoke Generator
   Appropriate Size
   Tinder
   Kindling
   Placement
   Aflame within 90 seconds

7. PLANTS & INSECTS (WSVX) M / NM
   Located 1/2 canteen cup of edible plant
   1/2 canteen cup of prepared charcoal available
   Consumed the prepared plant
   Continue to live off the land
   Consume a prepared insect

8. TOOLS & WEAPONS (MSV.2.11) M / NM

Bowl

Wood split
Bark stripped
Coal burned
4inch deep, 4 inch diameter
Does not leak

Simple Club

Hardwood used
Bark stripped
Fire hardened (if required)
Club functional
Rounded ends

9. TOOL OR WEAPON (MSV.2.11) M / NM
Hardwood used
Bark stripped
Fire hardened (if required)
Functional able
One of the three following will be made: Ice spud, Ice skimmer, Slingshot

10. FISH & GAME (MSV.2.10) M / NM
Dress and/or skin game
Prepared game for consumption
Consumed game
Hide fleshed, brained, and smoked
Hide sewn and suitable for intended uses

11. TRAPS & SNARES (MSV.2.9) M / NM
Employment tech. appropriate for intended animal
    a. Location
    b. Presentation
    c. Construction
Loop size and ground clearance correct
Bait used
Split stick if required

12. PATHGUARD (MSV.2.9) M / NM
Likely avenue of approach
Produces noise
Concealed
Appropriate tactical distance from the shelter

Made in the USA
Monee, IL
04 November 2026

67384906R00098